T0276849

Concepts and Applications of Continuum Mechanics

Concepts and Applications of Continuum Mechanics

Edited by **Derek Pearce**

New York

Published by NY Research Press,
23 West, 55th Street, Suite 816,
New York, NY 10019, USA
www.nyresearchpress.com

Concepts and Applications of Continuum Mechanics
Edited by Derek Pearce

International Standard Book Number: 978-1-63238-091-3 (Hardback)

Printed in the United States of America.

Contents

	Preface	VII
Chapter 1	**Spencer Operator and Applications:** **From Continuum Mechanics to Mathematical Physics** J.F. Pommaret	**1**
Chapter 2	**Continuum Mechanics of Solid Oxide Fuel Cells** **Using Three-Dimensional Reconstructed Microstructures** Sushrut Vaidya and Jeong-Ho Kim	**33**
Chapter 3	**Transversality Condition in Continuum Mechanics** Jianlin Liu	**49**
Chapter 4	**Incompressible Non-Newtonian Fluid Flows** Quoc-Hung Nguyen and Ngoc-Diep Nguyen	**63**
Chapter 5	**Noise and Vibration** **in Complex Hydraulic Tubing Systems** Chuan-Chiang Chen	**89**
Chapter 6	**Analysis Precision Machining Process** **Using Finite Element Method** Xuesong Han	**105**
Chapter 7	**Energy Dissipation Criteria** **for Surface Contact Damage Evaluation** Yong X. Gan	**123**
Chapter 8	**Progressive Stiffness Loss Analysis of Symmetric Laminated** **Plates due to Transverse Cracks Using the MLGFM** Roberto Dalledone Machado, Antonio Tassini Jr., Marcelo Pinto da Silva and Renato Barbieri	**139**
	Permissions	
	List of Contributors	

Preface

Covering every aspect of Continuum Mechanics, this book brilliantly elucidates its concepts and applications. Continuum mechanics is the base of Applied Mechanics. There are a number of books on Continuum Mechanics emphasizing on the macro-scale mechanical conduct of materials. Unlike traditional Continuum Mechanics books, this book provides synopsis on the developments in some specific areas of Continuum Mechanics. This book focuses primarily on the applications aspects. Energy materials and systems i.e. fuel cell materials and electrodes, substance deportation and mechanical response/deformation of plates, pipelines etc. have been covered under the applications described in this book. Researchers from different fields will benefit from reading the mechanics approach to solve engineering problems.

After months of intensive research and writing, this book is the end result of all who devoted their time and efforts in the initiation and progress of this book. It will surely be a source of reference in enhancing the required knowledge of the new developments in the area. During the course of developing this book, certain measures such as accuracy, authenticity and research focused analytical studies were given preference in order to produce a comprehensive book in the area of study.

This book would not have been possible without the efforts of the authors and the publisher. I extend my sincere thanks to them. Secondly, I express my gratitude to my family and well-wishers. And most importantly, I thank my students for constantly expressing their willingness and curiosity in enhancing their knowledge in the field, which encourages me to take up further research projects for the advancement of the area.

Editor

Spencer Operator and Applications: From Continuum Mechanics to Mathematical Physics

J.F. Pommaret

CERMICS, Ecole Nationale des Ponts et Chaussées,
France

1. Introduction

Let us revisit briefly the foundation of n-dimensional elasticity theory as it can be found today in any textbook, restricting our study to $n = 2$ for simplicity. If $x = (x^1, x^2)$ is a point in the plane and $\xi = (\xi^1(x), \xi^2(x))$ is the displacement vector, lowering the indices by means of the Euclidean metric, we may introduce the "small" deformation tensor $\epsilon = (\epsilon_{ij} = \epsilon_{ji} = (1/2)(\partial_i \xi_j + \partial_j \xi_i))$ with $n(n+1)/2 = 3$ (independent) *components* ($\epsilon_{11}, \epsilon_{12} = \epsilon_{21}, \epsilon_{22}$). If we study a part of a deformed body, for example a thin elastic plane sheet, by means of a variational principle, we may introduce the local density of free energy $\varphi(\epsilon) = \varphi(\epsilon_{ij} | i \leq j) = \varphi(\epsilon_{11}, \epsilon_{12}, \epsilon_{22})$ and vary the total free energy $F = \int \varphi(\epsilon) dx$ with $dx = dx^1 \wedge dx^2$ by introducing $\sigma^{ij} = \partial \varphi / \partial \epsilon_{ij}$ for $i \leq j$ in order to obtain $\delta F = \int (\sigma^{11} \delta \epsilon_{11} + \sigma^{12} \delta \epsilon_{12} + \sigma^{22} \delta \epsilon_{22}) dx$. Accordingly, the "decision" to define the stress tensor σ by a symmetric matrix with $\sigma^{12} = \sigma^{21}$ is purely artificial within such a variational principle. Indeed, the usual Cauchy device (1828) assumes that each element of a boundary surface is acted on by a surface density of force $\vec{\sigma}$ with a linear dependence $\vec{\sigma} = (\sigma^{ir}(x) n_r)$ on the outward normal unit vector $\vec{n} = (n_r)$ and does not make any assumption on the stress tensor. It is only by an equilibrium of forces and couples, namely the well known *phenomenological static torsor equilibrium*, that one can "prove" the symmetry of σ. However, even if we assume this symmetry, *we now need the different summation* $\sigma^{ij} \delta \epsilon_{ij} = \sigma^{11} \delta \epsilon_{11} + 2\sigma^{12} \delta \epsilon_{12} + \sigma^{22} \delta \epsilon_{22} = \sigma^{ir} \partial_r \delta \xi_i$. An integration by parts and a change of sign produce the volume integral $\int (\partial_r \sigma^{ir}) \delta \xi_i dx$ leading to the stress equations $\partial_r \sigma^{ir} = 0$. *The classical approach to elasticity theory, based on invariant theory with respect to the group of rigid motions, cannot therefore describe equilibrium of torsors by means of a variational principle where the proper torsor concept is totally lacking.*

There is another equivalent procedure dealing with a *variational calculus with constraint*. Indeed, as we shall see in Section 7, the deformation tensor is not any symmetric tensor as it must satisfy $n^2(n^2 - 1)/12$ compatibility conditions (CC), that is only $\partial_{22} \epsilon_{11} + \partial_{11} \epsilon_{22} - 2\partial_{12} \epsilon_{12} = 0$ when $n = 2$. In this case, introducing the *Lagrange multiplier* $-\phi$ for convenience, *we have to vary* $\int (\varphi(\epsilon) - \phi(\partial_{22}\epsilon_{11} + \partial_{11}\epsilon_{22} - 2\partial_{12}\epsilon_{12})) dx$ *for an arbitrary* ϵ. A double integration by parts now provides the parametrization $\sigma^{11} = \partial_{22}\phi, \sigma^{12} = \sigma^{21} = -\partial_{12}\phi, \sigma^{22} = \partial_{11}\phi$ of the stress equations by means of the Airy function ϕ and the *formal adjoint* of the CC, *on the condition to observe that we have in fact* $2\sigma^{12} = -2\partial_{12}\phi$ as another way to understand the deep meaning of the factor "2" in the summation. In arbitrary dimension, it just remains to notice that the above compatibility conditions are nothing else but the linearized Riemann tensor in Riemannan geometry, a crucial mathematical tool in the theory of general relativity.

It follows that the only possibility to revisit the foundations of engineering and mathematical physics is to use new mathematical methods, namely the theory of systems of partial differential equations and Lie pseudogroups developped by D.C. Spencer and coworkers during the period 1960-1975. In particular, Spencer invented the first order operator now wearing his name in order to bring in a canonical way the formal study of systems of ordinary differential (OD) or partial differential (PD) equations to that of equivalent first order systems. However, despite its importance, the *Spencer operator* is rarely used in mathematics today and, up to our knowledge, has never been used in engineering or mathematical physics. The main reason for such a situation is that the existing papers, largely based on hand-written lecture notes given by Spencer to his students (the author was among them in 1969) are quite technical and the problem also lies in the only "accessible" book "Lie equations" he published in 1972 with A. Kumpera. Indeed, the reader can easily check by himself that *the core of this book has nothing to do with its introduction* recalling known differential geometric concepts on which most of physics is based today.

The first and technical purpose of this chapter, an extended version of a lecture at the second workshop on Differential Equations by Algebraic Methods (DEAM2, february 9-11, 2011, Linz, Austria), is to recall briefly its definition, both in the framework of systems of linear ordinary or partial differential equations and in the framework of differential modules. The local theory of Lie pseudogroups and the corresponding non-linear framework are also presented for the first time in a rather elementary manner though it is a difficult task.

The second and central purpose is to prove that the use of the Spencer operator constitutes the *common secret* of the three following famous books published about at the same time in the beginning of the last century, though they do not seem to have anything in common at first sight as they are successively dealing with the foundations of elasticity theory, commutative algebra, electromagnetism (EM) and general relativity (GR):

[C] E. and F. COSSERAT: "Théorie des Corps Déformables", Hermann, Paris, 1909.
[M] F.S. MACAULAY: "The Algebraic Theory of Modular Systems", Cambridge, 1916.
[W] H. WEYL: "Space, Time, Matter", Springer, Berlin, 1918 (1922, 1958; Dover, 1952).

Meanwhile we shall point out the striking importance of the second book for studying *identifiability* in control theory. We shall also obtain from the previous results the group theoretical unification of finite elements in engineering sciences (elasticity, heat, electromagnetism), solving the *torsor problem* and recovering in a purely mathematical way known *field-matter coupling phenomena* (piezzoelectricity, photoelasticity, streaming birefringence, viscosity, ...).

As a byproduct and though disturbing it may be, the third and perhaps essential purpose is to prove that *these unavoidable new differential and homological methods contradict the existing mathematical foundations of both engineering (continuum mechanics, electromagnetism) and mathematical (gauge theory, general relativity) physics.*

Many explicit examples will illustate this chapter which is deliberately written in a rather self-contained way to be accessible to a large audience, which does not mean that it is elementary in view of the number of new concepts that must be patched together. However, the reader must never forget that *each formula* appearing in this new general framework has been used explicitly or implicitly in [C], [M] and [W] for a mechanical, mathematical or physical purpose.

2. From Lie groups to Lie pseudogroups

Evariste Galois (1811-1832) introduced the word *"group"* for the first time in 1830. Then the group concept slowly passed from algebra (groups of permutations) to geometry (groups of transformations). It is only in 1880 that Sophus Lie (1842-1899) studied the groups of transformations depending on a finite number of parameters and now called *Lie groups of transformations*. Let X be a manifold with local coordinates $x = (x^1, ..., x^n)$ and G be a *Lie group*, that is another manifold with local coordinates $a = (a^1, ..., a^p)$ called *parameters* with a composition $G \times G \to G : (a, b) \to ab$, an *inverse* $G \to G : a \to a^{-1}$ and an *identity* $e \in G$ satisfying:

$$(ab)c = a(bc) = abc, \qquad aa^{-1} = a^{-1}a = e, \qquad ae = ea = a, \qquad \forall a, b, c \in G$$

Definition 2.1. *G is said to act on X if there is a map $X \times G \to X : (x, a) \to y = ax = f(x, a)$ such that $(ab)x = a(bx) = abx, \forall a, b \in G, \forall x \in X$ and, for simplifying the notations, we shall use global notations even if only local actions are existing. The set $G_x = \{a \in G \mid ax = x\}$ is called the isotropy subgroup of G at $x \in X$. The action is said to be effective if $ax = x, \forall x \in X \Rightarrow a = e$. A subset $S \subset X$ is said to be invariant under the action of G if $aS \subset S, \forall a \in G$ and the orbit of $x \in X$ is the invariant subset $Gx = \{ax \mid a \in G\} \subset X$. If G acts on two manifolds X and Y, a map $f : X \to Y$ is said to be equivariant if $f(ax) = af(x), \forall x \in X, \forall a \in G$.*

For reasons that will become clear later on, it is often convenient to introduce the *graph* $X \times G \to X \times X : (x, a) \to (x, y = ax)$ of the action. In the product $X \times X$, the first factor is called the *source* while the second factor is called the *target*.

Definition 2.2. *The action is said to be free if the graph is injective and transitive if the graph is surjective. The action is said to be simply transitive if the graph is an isomorphism and X is said to be a principal homogeneous space (PHS) for G.*

In order to fix the notations, we quote without any proof the *"Three Fundamental Theorems of Lie"* that will be of constant use in the sequel ([26]):

First fundamental theorem: The orbits $x = f(x_0, a)$ satisfy the system of PD equations $\partial x^i / \partial a^\sigma = \theta^i_\rho(x) \omega^\rho_\sigma(a)$ with $det(\omega) \neq 0$. The vector fields $\theta_\rho = \theta^i_\rho(x) \partial_i$ are called *infinitesimal generators* of the action and are linearly independent over the constants when the action is effective.

If X is a manifold, we denote as usual by $T = T(X)$ the *tangent bundle* of X, by $T^* = T^*(X)$ the *cotangent bundle*, by $\wedge^r T^*$ the *bundle of r-forms* and by $S_q T^*$ the *bundle of q-symmetric tensors*. More generally, let $\mathcal{E}$ be a *fibered manifold*, that is a manifold with local coordinates (x^i, y^k) for $i = 1, ..., n$ and $k = 1, ..., m$ simply denoted by (x, y), *projection* $\pi : \mathcal{E} \to X : (x, y) \to (x)$ and changes of local coordinates $\bar{x} = \varphi(x), \bar{y} = \psi(x, y)$. If $\mathcal{E}$ and $\mathcal{F}$ are two fibered manifolds over X with respective local coordinates (x, y) and (x, z), we denote by $\mathcal{E} \times_X \mathcal{F}$ the *fibered product* of $\mathcal{E}$ and $\mathcal{F}$ over X as the new fibered manifold over X with local coordinates (x, y, z). We denote by $f : X \to \mathcal{E} : (x) \to (x, y = f(x))$ a global *section* of $\mathcal{E}$, that is a map such that $\pi \circ f = id_X$ but local sections over an open set $U \subset X$ may also be considered when needed. Under a change of coordinates, a section transforms like $\bar{f}(\varphi(x)) = \psi(x, f(x))$ and the derivatives transform like:

$$\frac{\partial \bar{f}^l}{\partial \bar{x}^r}(\varphi(x)) \partial_i \varphi^r(x) = \frac{\partial \psi^l}{\partial x^i}(x, f(x)) + \frac{\partial \psi^l}{\partial y^k}(x, f(x)) \partial_i f^k(x)$$

We may introduce new coordinates (x^i, y^k, y^k_i) transforming like:

$$\bar{y}^l_r \partial_i \varphi^r(x) = \frac{\partial \psi^l}{\partial x^i}(x, y) + \frac{\partial \psi^l}{\partial y^k}(x, y) y^k_i$$

We shall denote by $J_q(\mathcal{E})$ the q-jet bundle of $\mathcal{E}$ with local coordinates $(x^i, y^k, y^k_i, y^k_{ij}, ...) = (x, y_q)$ called jet coordinates and sections $f_q : (x) \rightarrow (x, f^k(x), f^k_i(x), f^k_{ij}(x), ...) = (x, f_q(x))$ transforming like the sections $j_q(f) : (x) \rightarrow (x, f^k(x), \partial_i f^k(x), \partial_{ij} f^k(x), ...) = (x, j_q(f)(x))$ where both f_q and $j_q(f)$ are over the section f of $\mathcal{E}$. Of course $J_q(\mathcal{E})$ is a fibered manifold over X with projection π_q while $J_{q+r}(\mathcal{E})$ is a fibered manifold over $J_q(\mathcal{E})$ with projection $\pi^{q+r}_q, \forall r \geq 0$.

Definition 2.3. A system of order q on $\mathcal{E}$ is a fibered submanifold $\mathcal{R}_q \subset J_q(\mathcal{E})$ and a solution of $\mathcal{R}_q$ is a section f of $\mathcal{E}$ such that $j_q(f)$ is a section of $\mathcal{R}_q$.

Definition 2.4. When the changes of coordinates have the linear form $\bar{x} = \varphi(x), \bar{y} = A(x)y$, we say that $\mathcal{E}$ is a vector bundle over X and denote for simplicity a vector bundle and its set of sections by the same capital letter E. When the changes of coordinates have the form $\bar{x} = \varphi(x), \bar{y} = A(x)y + B(x)$ we say that $\mathcal{E}$ is an affine bundle over X and we define the associated vector bundle E over X by the local coordinates (x, v) changing like $\bar{x} = \varphi(x), \bar{v} = A(x)v$.

Definition 2.5. If the tangent bundle $T(\mathcal{E})$ has local coordinates (x, y, u, v) changing like $\bar{u}^j = \partial_i \varphi^j(x) u^i, \bar{v}^l = \frac{\partial \psi^l}{\partial x^i}(x, y) u^i + \frac{\partial \psi^l}{\partial y^k}(x, y) v^k$, we may introduce the vertical bundle $V(\mathcal{E}) \subset T(\mathcal{E})$ as a vector bundle over $\mathcal{E}$ with local coordinates (x, y, v) obtained by setting $u = 0$ and changes $\bar{v}^l = \frac{\partial \psi^l}{\partial y^k}(x, y) v^k$. Of course, when $\mathcal{E}$ is an affine bundle with associated vector bundle E over X, we have $V(\mathcal{E}) = \mathcal{E} \times_X E$.

For a later use, if $\mathcal{E}$ is a fibered manifold over X and f is a section of $\mathcal{E}$, we denote by $f^{-1}(V(\mathcal{E}))$ the reciprocal image of $V(\mathcal{E})$ by f as the vector bundle over X obtained when replacing (x, y, v) by $(x, f(x), v)$ in each chart. It is important to notice in variational calculus that a variation δf of f is such that $\delta f(x)$, as a vertical vector field not necessary "small", is a section of this vector bundle and that $(f, \delta f)$ is nothing else than a section of $V(\mathcal{E})$ over X.

We now recall a few basic geometric concepts that will be constantly used. First of all, if $\xi, \eta \in T$, we define their bracket $[\xi, \eta] \in T$ by the local formula $([\xi, \eta])^i(x) = \xi^r(x) \partial_r \eta^i(x) - \eta^s(x) \partial_s \xi^i(x)$ leading to the Jacobi identity $[\xi, [\eta, \zeta]] + [\eta, [\zeta, \xi]] + [\zeta, [\xi, \eta]] = 0, \forall \xi, \eta, \zeta \in T$ allowing to define a Lie algebra and to the useful formula $[T(f)(\xi), T(f)(\eta)] = T(f)([\xi, \eta])$ where $T(f) : T(X) \rightarrow T(Y)$ is the tangent mapping of a map $f : X \rightarrow Y$.

Second fundamental theorem: If $\theta_1, ..., \theta_p$ are the infinitesimal generators of the effective action of a lie group G on X, then $[\theta_\rho, \theta_\sigma] = c^\tau_{\rho\sigma} \theta_\tau$ where the $c^\tau_{\rho\sigma}$ are the structure constants of a Lie algebra of vector fields which can be identified with $\mathcal{G} = T_e(G)$.

When $I = \{i_1 < ... < i_r\}$ is a multi-index, we may set $dx^I = dx^{i_1} \wedge ... \wedge dx^{i_r}$ for describing $\wedge^r T^*$ and introduce the exterior derivative $d : \wedge^r T^* \rightarrow \wedge^{r+1} T^* : \omega = \omega_I dx^I \rightarrow d\omega = \partial_i \omega_I dx^i \wedge dx^I$ with $d^2 = d \circ d \equiv 0$ in the Poincaré sequence:

$$\wedge^0 T^* \xrightarrow{d} \wedge^1 T^* \xrightarrow{d} \wedge^2 T^* \xrightarrow{d} ... \xrightarrow{d} \wedge^n T^* \longrightarrow 0$$

The *Lie derivative* of an r-form with respect to a vector field $\xi \in T$ is the linear first order operator $\mathcal{L}(\xi)$ linearly depending on $j_1(\xi)$ and uniquely defined by the following three properties:

1. $\mathcal{L}(\xi)f = \xi.f = \xi^i \partial_i f, \forall f \in \wedge^0 T^* = C^\infty(X)$.
2. $\mathcal{L}(\xi)d = d\mathcal{L}(\xi)$.
3. $\mathcal{L}(\xi)(\alpha \wedge \beta) = (\mathcal{L}(\xi)\alpha) \wedge \beta + \alpha \wedge (\mathcal{L}(\xi)\beta), \forall \alpha, \beta \in \wedge T^*$.

It can be proved that $\mathcal{L}(\xi) = i(\xi)d + di(\xi)$ where $i(\xi)$ is the *interior multiplication* $(i(\xi)\omega)_{i_1...i_r} = \xi^i \omega_{ii_1...i_r}$ and that $[\mathcal{L}(\xi), \mathcal{L}(\eta)] = \mathcal{L}(\xi) \circ \mathcal{L}(\eta) - \mathcal{L}(\eta) \circ \mathcal{L}(\xi) = \mathcal{L}([\xi, \eta]), \forall \xi, \eta \in T$.

Using crossed-derivatives in the PD equations of the First Fundamental Theorem and introducing the family of 1-forms $\omega^\tau = \omega^\tau_\sigma(a)da^\sigma$ both with the matrix $\alpha = \omega^{-1}$ of right invariant vector fields, we obtain the *compatibility conditions* (CC) expressed by the following corollary where one must care about the sign used:

Corollary 2.1. *One has the Maurer-Cartan (MC) equations* $d\omega^\tau + c^\tau_{\rho\sigma}\omega^\rho \wedge \omega^\sigma = 0$ *or the equivalent relations* $[\alpha_\rho, \alpha_\sigma] = c^\tau_{\rho\sigma}\alpha_\tau$.

Applying d to the MC equations and substituting, we obtain the *integrability conditions* (IC):

Third fundamental theorem For any Lie algebra $\mathcal{G}$ defined by structure constants satisfying :

$$c^\tau_{\rho\sigma} + c^\tau_{\sigma\rho} = 0, \qquad c^\lambda_{\mu\rho}c^\mu_{\sigma\tau} + c^\lambda_{\mu\sigma}c^\mu_{\tau\rho} + c^\lambda_{\mu\tau}c^\mu_{\rho\sigma} = 0$$

one can construct an analytic group G such that $\mathcal{G} = T_e(G)$.

Example 2.1. *Considering the affine group of transformations of the real line* $y = a^1 x + a^2$, *we obtain* $\theta_1 = x\partial_x, \theta_2 = \partial_x \Rightarrow [\theta_1, \theta_2] = -\theta_2$ *and thus* $\omega^1 = (1/a^1)da^1, \omega^2 = da^2 - (a^2/a^1)da^1 \Rightarrow d\omega^1 = 0, d\omega^2 - \omega^1 \wedge \omega^2 = 0 \Leftrightarrow [\alpha_1, \alpha_2] = -\alpha_2$ *with* $\alpha_1 = a^1\partial_1 + a^2\partial_2, \alpha_2 = \partial_2$.

Only ten years later Lie discovered that the Lie groups of transformations are only particular examples of a wider class of groups of transformations along the following definition where $aut(X)$ denotes the group of all local diffeomorphisms of X:

Definition 2.6. *A Lie pseudogroup of transformations* $\Gamma \subset aut(X)$ *is a group of transformations solutions of a system of OD or PD equations such that, if* $y = f(x)$ *and* $z = g(y)$ *are two solutions, called finite transformations, that can be composed, then* $z = g \circ f(x) = h(x)$ *and* $x = f^{-1}(y) = g(y)$ *are also solutions while* $y = x$ *is a solution.*

The underlying system may be nonlinear and of high order as we shall see later on. We shall speak of an *algebraic pseudogroup* when the system is defined by *differential polynomials* that is polynomials in the derivatives. In the case of Lie groups of transformations the system is obtained by differentiating the action law $y = f(x, a)$ with respect to x as many times as necessary in order to eliminate the parameters. Looking for transformations "close" to the identity, that is setting $y = x + t\xi(x) + ...$ when $t \ll 1$ is a small constant parameter and passing to the limit $t \to 0$, we may linearize the above nonlinear *system of finite Lie equations* in order to obtain a linear *system of infinitesimal Lie equations* of the same order for vector fields. Such a system has the property that, if ξ, η are two solutions, then $[\xi, \eta]$ is also a solution. Accordingly, the set $\Theta \subset T$ of solutions of this new system satifies $[\Theta, \Theta] \subset \Theta$ and can therefore be considered as the Lie algebra of Γ.

Though the collected works of Lie have been published by his student F. Engel at the end of the 19th century, these ideas did not attract a large audience because the fashion in Europe was analysis. Accordingly, at the beginning of the 20th century and for more than fifty years, only two frenchmen tried to go further in the direction pioneered by Lie, namely Elie Cartan (1869-1951) who is quite famous today and Ernest Vessiot (1865-1952) who is almost ignored today, each one deliberately ignoring the other during his life for a precise reason that we now explain with details as it will surprisingly constitute the heart of this chapter. (The author is indebted to Prof. Maurice Janet (1888-1983), who was a personal friend of Vessiot, for the many documents he gave him, partly published in [25]). Roughly, the idea of many people at that time was to extend the work of Galois along the following scheme of increasing difficulty:

1) *Galois theory* ([34]): Algebraic equations and permutation groups.
2) *Picard-Vessiot theory* ([17]): OD equations and Lie groups.
3) *Differential Galois theory* ([24],[37]): PD equations and Lie pseudogroups.

In 1898 Jules Drach (1871-1941) got and published a thesis ([9]) with a jury made by Gaston Darboux, Emile Picard and Henri Poincaré, the best leading mathematicians of that time. However, despite the fact that it contains ideas quite in advance on his time such as the concept of a "differential field" only introduced by J.-F. Ritt in 1930 ([31]), the jury did not notice that the main central result was wrong: Cartan found the counterexamples, Vessiot recognized the mistake, Paul Painlevé told it to Picard who agreed but Drach never wanted to acknowledge this fact and was supported by the influent Emile Borel. As a byproduct, everybody flew out of this "affair", never touching again the Galois theory. After publishing a prize-winning paper in 1904 where he discovered for the first time that the differential Galois theory must be a theory of (irreducible) PHS for algebraic pseudogroups, Vessiot remained alone, trying during thirty years to correct the mistake of Drach that we finally corrected in 1983 ([24]).

3. Cartan versus Vessiot : The structure equations

We study first the work of Cartan which can be divided into two parts. The first part, for which he invented exterior calculus, may be considered as a tentative to extend the MC equations from Lie groups to Lie pseudogroups. The idea for that is to consider the system of order q and its *prolongations* obtained by differentiating the equations as a way to know certain derivatives called *principal* from all the other arbitrary ones called *parametric* in the sense of Janet ([13]). Replacing the derivatives by jet coordinates, we may try to copy the procedure leading to the MC equations by using a kind of "composition" and "inverse" on the jet coordinates. For example, coming back to the last definition, we get successively:

$$\frac{\partial h}{\partial x} = \frac{\partial g}{\partial y}\frac{\partial f}{\partial x}, \qquad \frac{\partial^2 h}{\partial x^2} = \frac{\partial^2 g}{\partial y^2}\frac{\partial f}{\partial x}\frac{\partial f}{\partial x} + \frac{\partial g}{\partial y}\frac{\partial^2 f}{\partial x^2}, \dots$$

Now if $g = f^{-1}$ then $g \circ f = id$ and thus $\frac{\partial g}{\partial y}\frac{\partial f}{\partial x} = 1, \dots$ while the new identity $id_q = j_q(id)$ is made by the successive derivatives of $y = x$, namely $(1, 0, 0, \dots)$. These *awfully complicated computations* bring a lot of structure constants and have been so much superseded by the work of Donald C. Spencer (1912-2001) ([11],[12],[18],[33]) that, in our opinion based on thirty years of explicit computations, this tentative has only been used for classification problems and is not useful for applications compared to the results of the next sections. In a single concluding sentence, *Cartan has not been able to "go down" to the base manifold X while Spencer did succeed fifty years later.*

We shall now describe the second part with more details as it has been (and still is !) the crucial tool used in both engineering (analytical and continuum mechanics) and mathematical (gauge theory and general relativity) physics in an absolutely contradictory manner. We shall try to use the least amount of mathematics in order to prepare the reader for the results presented in the next sections. For this let us start with an elementary experiment that will link at once continuum mechanics and gauge theory in an unusual way. Let us put a thin elastic rectilinear rubber band along the x axis on a flat table and translate it along itself. The band will remain identical as no deformation can be produced by this constant translation. However, if we move each point continuously along the same direction but in a point depending way, for example fixing one end and pulling on the other end, there will be of course a deformation of the elastic band according to the Hooke law. It remains to notice that a constant translation can be written in the form $y = x + a^2$ as in Example 2.1 while a point depending translation can be written in the form $y = x + a^2(x)$ which is written in any textbook of continuum mechanics in the form $y = x + \xi(x)$ by introducing the *displacement vector* ξ. However nobody could even imagine to make a^1 also point depending and to consider $y = a^1(x)x + a^2(x)$ as we shall do in Example 7.2.We also notice that the movement of a rigid body in space may be written in the form $y = a(t)x + b(t)$ where now $a(t)$ is a time depending orthogonal matrix and $b(t)$ is a time depending vector. What makes all the difference between the two examples is that the group is *acting* on x in the first but *not acting* on t in the second. Finally, a point depending rotation or dilatation is not accessible to intuition and the general theory must be done in the following manner.

If X is a manifold and G is a lie group *not acting necessarily* on X, let us consider maps $a : X \to G : (x) \to (a(x))$ or equivalently sections of the trivial (principal) bundle $X \times G$ over X. If $x + dx$ is a point of X close to x, then $T(a)$ will provide a point $a + da = a + \frac{\partial a}{\partial x} dx$ close to a on G. We may bring a back to e on G by acting on a with a^{-1}, *either on the left or on the right*, getting therefore a 1-form $a^{-1}da = A$ or $daa^{-1} = B$. As $aa^{-1} = e$ we also get $daa^{-1} = -ada^{-1} = -b^{-1}db$ if we set $b = a^{-1}$ as a way to link A with B. When there is an action $y = ax$, we have $x = a^{-1}y = by$ and thus $dy = dax = daa^{-1}y$, a result leading through the First Fundamental Theorem of Lie to the equivalent formulas:

$$a^{-1}da = A = (A_i^\tau(x)dx^i = -\omega_\sigma^\tau(b(x))\partial_i b^\sigma(x)dx^i)$$

$$daa^{-1} = B = (B_i^\tau(x)dx^i = \omega_\sigma^\tau(a(x))\partial_i a^\sigma(x)dx^i)$$

Introducing the induced bracket $[A, A](\xi, \eta) = [A(\xi), A(\eta)] \in \mathcal{G}, \forall \xi, \eta \in T$, we may define the 2-form $dA - [A, A] = F \in \wedge^2 T^* \otimes \mathcal{G}$ by the local formula (care to the sign):

$$\partial_i A_j^\tau(x) - \partial_j A_i^\tau(x) - c_{\rho\sigma}^\tau A_i^\rho(x)A_j^\sigma(x) = F_{ij}^\tau(x)$$

and obtain from the second fundamental theorem:

Theorem 3.1. *There is a nonlinear gauge sequence:*

$$X \times G \longrightarrow T^* \otimes \mathcal{G} \xrightarrow{\text{MC}} \wedge^2 T^* \otimes \mathcal{G}$$
$$a \longrightarrow a^{-1}da = A \longrightarrow dA - [A, A] = F$$

Choosing a "close" to e, that is $a(x) = e + t\lambda(x) + ...$ and linearizing as usual, we obtain the linear operator $d : \wedge^0 T^* \otimes \mathcal{G} \to \wedge^1 T^* \otimes \mathcal{G} : (\lambda^\tau(x)) \to (\partial_i \lambda^\tau(x))$ leading to:

Corollary 3.1. *There is a linear gauge sequence:*

$$\wedge^0 T^* \otimes \mathcal{G} \xrightarrow{d} \wedge^1 T^* \otimes \mathcal{G} \xrightarrow{d} \wedge^2 T^* \otimes \mathcal{G} \xrightarrow{d} ... \xrightarrow{d} \wedge^n T^* \otimes \mathcal{G} \longrightarrow 0$$

which is the tensor product by $\mathcal{G}$ of the Poincaré sequence:

Remark 3.1. *When the physicists C.N. Yang and R.L. Mills created (non-abelian) gauge theory in 1954 ([38],[39]), their work was based on these results which were the only ones known at that time, the best mathematical reference being the well known book by S. Kobayashi and K. Nomizu ([15]). It follows that the only possibility to describe elecromagnetism (EM) within this framework was to call A the Yang-Mills potential and F the Yang-Mills field (a reason for choosing such notations) on the condition to have $dim(G) = 1$ in the abelian situation $c = 0$ and to use a Lagrangian depending on $F = dA - [A, A]$ in the general case. Accordingly the idea was to select the unitary group $U(1)$, namely the unit circle in the complex plane with Lie algebra the tangent line to this circle at the unity $(1, 0)$. It is however important to notice that the resulting Maxwell equations $dF = 0$ have no equivalent in the non-abelian case $c \neq 0$.*

Just before Albert Einstein visited Paris in 1922, Cartan published many short Notes ([5]) announcing long papers ([6]) where he selected G to be the Lie group involved in the Poincaré (conformal) group of space-time preserving (up to a function factor) the Minkowski metric $\omega = (dx^1)^2 + (dx^2)^2 + (dx^3)^2 - (dx^4)^2$ with $x^4 = ct$ where c is the speed of light. In the first case F is decomposed into two parts, the *torsion* as a 2-form with value in translations on one side and the *curvature* as a 2-form with value in rotations on the other side. This result was looking coherent *at first sight* with the Hilbert variational scheme of general relativity (GR) introduced by Einstein in 1915 ([21],[38]) and leading to a Lagrangian depending on $F = dA - [A, A]$ as in the last remark.

In the meantime, Poincaré developped an invariant variational calculus ([22]) which has been used again without any quotation, successively by G. Birkhoff and V. Arnold (compare [4], 205-216 with [2], 326, Th 2.1). A particular case is well known by any student in the analytical mechanics of rigid bodies. Indeed, using standard notations, the movement of a rigid body is described in a fixed Cartesian frame by the formula $x(t) = a(t)x_0 + b(t)$ where $a(t)$ is a 3×3 time dependent orthogonal matrix (rotation) and $b(t)$ a time depending vector (translation) as we already said. Differentiating with respect to time by using a dot, the *absolute speed* is $v = \dot{x}(t) = \dot{a}(t)x_0 + \dot{b}(t)$ and we obtain the *relative speed* $a^{-1}(t)v = a^{-1}(t)\dot{a}(t)x_0 + a^{-1}(t)\dot{b}(t)$ by projection in a frame fixed in the body. Having in mind Example 2.1, it must be noticed that the so-called *Eulerian speed* $v = v(x, t) = \dot{a}a^{-1}x + \dot{b} - \dot{a}a^{-1}b$ only depends on the 1-form $B = (\dot{a}a^{-1}, \dot{b} - \dot{a}a^{-1}b)$. The Lagrangian (kinetic energy in this case) is thus a quadratic function of the 1-form $A = (a^{-1}\dot{a}, a^{-1}\dot{b})$ where $a^{-1}\dot{a}$ is a 3×3 skew symmetric time depending matrix. Hence, "surprisingly", this result is not coherent at all with EM where the Lagrangian is the quadratic expression $(\epsilon/2)E^2 - (1/2\mu)B^2$ because the electric field $\vec{E}$ and the magnetic field $\vec{B}$ are combined in the EM field F as a 2-form satisfying the first set of Maxwell equations $dF = 0$. The dielectric constant ϵ and the magnetic constant μ are leading to the electric induction $\vec{D} = \epsilon\vec{E}$ and the magnetic induction $\vec{H} = (1/\mu)\vec{B}$ in the second set of Maxwell equations. In view of the existence of well known field-matter couplings such as piezoelectricity and photoelasticity that will be described later on, such a situation is contradictory as it should lead to put on equal footing 1-forms and 2-forms contrary to any unifying mathematical scheme but no other substitute could have been provided at that time.

Let us now turn to the other way proposed by Vessiot in 1903 ([36]) and 1904 ([37]). Our purpose is only to sketch the main results that we have obtained in many books ([23-26], we do not know other references) and to illustrate them by a series of specific examples, asking the reader to imagine any link with what has been said.

1. If $\mathcal{E} = X \times X$, we shall denote by $\Pi_q = \Pi_q(X, X)$ the open subfibered manifold of $J_q(X \times X)$ defined independently of the coordinate system by $det(y_i^k) \neq 0$ with *source projection* $\alpha_q : \Pi_q \to X : (x, y_q) \to (x)$ and *target projection* $\beta_q : \Pi_q \to X : (x, y_q) \to (y)$. We shall sometimes introduce a copy Y of X with local coordinates (y) in order to avoid any confusion between the source and the target manifolds. Let us start with a Lie pseudogroup $\Gamma \subset aut(X)$ defined by a system $\mathcal{R}_q \subset \Pi_q$ of order q. In all the sequel we shall suppose that the system is involutive (see next section) and that Γ is *transitive* that is $\forall x, y \in X, \exists f \in \Gamma, y = f(x)$ or, equivalently, the map $(\alpha_q, \beta_q) : \mathcal{R}_q \to X \times X : (x, y_q) \to (x, y)$ is surjective.

2. The Lie algebra $\Theta \subset T$ of infinitesimal transformations is then obtained by linearization, setting $y = x + t\xi(x) + \ldots$ and passing to the limit $t \to 0$ in order to obtain the linear involutive system $R_q = id_q^{-1}(V(\mathcal{R}_q)) \subset J_q(T)$ by reciprocal image with $\Theta = \{\xi \in T | j_q(\xi) \in R_q\}$.

3. Passing from source to target, we may *prolong* the vertical infinitesimal transformations $\eta = \eta^k(y)\frac{\partial}{\partial y^k}$ to the jet coordinates up to order q in order to obtain:

$$\eta^k(y)\frac{\partial}{\partial y^k} + \frac{\partial \eta^k}{\partial y^r}y_i^r\frac{\partial}{\partial y_i^k} + \left(\frac{\partial^2 \eta^k}{\partial y^r \partial y^s}y_i^r y_j^s + \frac{\partial \eta^k}{\partial y^r}y_{ij}^r\right)\frac{\partial}{\partial y_{ij}^k} + \ldots$$

where we have replaced $j_q(f)(x)$ by y_q, each component beeing the "formal" derivative of the previous one .

4. As $[\Theta, \Theta] \subset \Theta$, we may use the Frobenius theorem in order to find a generating fundamental set of *differential invariants* $\{\Phi^\tau(y_q)\}$ up to order q which are such that $\Phi^\tau(\bar{y}_q) = \Phi^\tau(y_q)$ by using the chain rule for derivatives whenever $\bar{y} = g(y) \in \Gamma$ acting now on Y. Of course, in actual practice *one must use sections of R_q instead of solutions* but it is only in section 6 that we shall see why the use of the Spencer operator will be crucial for this purpose. Specializing the Φ^τ at $id_q(x)$ we obtain the *Lie form* $\Phi^\tau(y_q) = \omega^\tau(x)$ of $\mathcal{R}_q$.

5. The main discovery of Vessiot, fifty years in advance, has been to notice that the prolongation at order q of any horizontal vector field $\xi = \xi^i(x)\frac{\partial}{\partial x^i}$ commutes with the prolongation at order q of any vertical vector field $\eta = \eta^k(y)\frac{\partial}{\partial y^k}$, exchanging therefore the differential invariants. Keeping in mind the well known property of the Jacobian determinant while passing to the finite point of view, any (local) transformation $y = f(x)$ can be lifted to a (local) transformation of the differential invariants between themselves of the form $u \to \lambda(u, j_q(f)(x))$ allowing to introduce a *natural bundle* $\mathcal{F}$ over X by patching changes of coordinates $\bar{x} = \varphi(x), \bar{u} = \lambda(u, j_q(\varphi)(x))$. A section ω of $\mathcal{F}$ is called a *geometric object* or *structure* on X and transforms like $\bar{\omega}(f(x)) = \lambda(\omega(x), j_q(f)(x))$ or simply $\bar{\omega} = j_q(f)(\omega)$. This is a way to generalize vectors and tensors $(q = 1)$ or even connections $(q = 2)$. As a byproduct we have $\Gamma = \{f \in aut(X) | \Phi_\omega(j_q(f)) = j_q(f)^{-1}(\omega) = \omega\}$ as a new way to write out the Lie form and we may say that Γ *preserves* ω. We also obtain $\mathcal{R}_q = \{f_q \in \Pi_q | f_q^{-1}(\omega) = \omega\}$. Coming back to the infinitesimal point of view and setting $f_t = exp(t\xi) \in aut(X), \forall \xi \in T$, we may define the *ordinary Lie derivative* with value in

$\omega^{-1}(V(\mathcal{F}))$ by the formula :

$$\mathcal{D}\xi = \mathcal{D}_\omega \xi = \mathcal{L}(\xi)\omega = \frac{d}{dt} j_q(f_t)^{-1}(\omega)|_{t=0} \Rightarrow \Theta = \{\xi \in T | \mathcal{L}(\xi)\omega = 0\}$$

while we have $x \to x + t\xi(x) + ... \Rightarrow u^\tau \to u^\tau + t\partial_\mu \xi^k L_k^{\tau\mu}(u) + ...$ where $\mu = (\mu_1, ..., \mu_n)$ is a multi-index as a way to write down the system of infinitesimal Lie equations in the *Medolaghi form*:

$$\Omega^\tau \equiv (\mathcal{L}(\xi)\omega)^\tau \equiv -L_k^{\tau\mu}(\omega(x))\partial_\mu \xi^k + \xi^r \partial_r \omega^\tau(x) = 0$$

6. By analogy with "special" and "general" relativity, we shall call the given section *special* and any other arbitrary section *general*. The problem is now to study the formal properties of the linear system just obtained with coefficients only depending on $j_1(\omega)$, exactly like L.P. Eisenhart did for $\mathcal{F} = S_2 T^*$ when finding the constant Riemann curvature condition for a metric ω with $det(\omega) \neq 0$ ([26], Example 10, p 249). Indeed, if any expression involving ω and its derivatives is a scalar object, it must reduce to a constant because Γ is assumed to be transitive and thus cannot be defined by any zero order equation. Now one can prove that the CC for $\bar{\omega}$, thus for ω too, only depend on the Φ and take the quasi-linear symbolic form $v \equiv I(u_1) \equiv A(u)u_x + B(u) = 0$, allowing to define an affine subfibered manifold $\mathcal{B}_1 \subset J_1(\mathcal{F})$ over $\mathcal{F}$. Now, if one has two sections ω and $\bar{\omega}$ of $\mathcal{F}$, the *equivalence problem* is to look for $f \in aut(X)$ such that $j_q(f)^{-1}(\omega) = \bar{\omega}$. When the two sections satisfy the same CC, the problem is sometimes locally possible (Lie groups of transformations, Darboux problem in analytical mechanics,...) but sometimes not ([23], p 333).

7. Instead of the CC for the equivalence problem, let us look for the *integrability conditions* (IC) for the system of infinitesimal Lie equations and suppose that, for the given section, all the equations of order $q + r$ are obtained by differentiating r times only the equations of order q, then it was claimed by Vessiot ([36] with no proof, see [26], p 209) that such a property is held if and only if there is an equivariant section $c : \mathcal{F} \to \mathcal{F}_1 : (x, u) \to (x, u, v = c(u))$ where $\mathcal{F}_1 = J_1(\mathcal{F})/\mathcal{B}_1$ is a natural vector bundle over $\mathcal{F}$ with local coordinates (x, u, v). Moreover, any such equivariant section depends on a finite number of constants c called *structure constants* and the IC for the *Vessiot structure equations* $I(u_1) = c(u)$ are of a polynomial form $J(c) = 0$.

8. Finally, when Y is no longer a copy of X, a system $\mathcal{A}_q \subset J_q(X \times Y)$ is said to be an *automorphic system* for a Lie pseudogroup $\Gamma \subset aut(Y)$ if, whenever $y = f(x)$ and $\bar{y} = \bar{f}(x)$ are two solutions, then there exists one and only one transformation $\bar{y} = g(y) \in \Gamma$ such that $\bar{f} = g \circ f$. Explicit tests for checking such a property formally have been given in [24] and can be implemented on computer in the differential algebraic framework.

Example 3.1. *(Principal homogeneous structure) When Γ is made by the translations $y^i = x^i + a^i$, the Lie form is $\Phi_i^k(y_1) \equiv y_i^k = \delta_i^k$ (Kronecker symbol) and the linearization is $\partial_i \xi^k = 0$. The natural bundle is $\mathcal{F} = T^* \times_X ... \times_X T^*$ (n times) with $det(\omega) \neq 0$ and the general Medolaghi form is $\omega_r^\tau \partial_i \xi^r + \xi^r \partial_r \omega_i^\tau = 0 \Leftrightarrow [\xi, \alpha_\tau] = 0$ with $\tau = 1, ..., n$ if $\alpha = (\alpha_\tau^i) = \omega^{-1}$. Using crossed derivatives, one finally gets the zero order equations:*

$$\xi^r \partial_r(\alpha_\rho^i(x)\alpha_\sigma^j(x)(\partial_i \omega_j^\tau(x) - \partial_j \omega_i^\tau(x))) = 0$$

leading therefore (up to sign) to the $n^2(n-1)/2$ Vessiot structure equations:

$$\partial_i \omega_j^\tau(x) - \partial_j \omega_i^\tau(x) = c_{\rho\sigma}^\tau \omega_i^\rho(x)\omega_j^\sigma(x)$$

This result proves that the MC equations are only examples of the Vessiot structure equations. We finally explain the name given to this structure ([26], p 296). Indeed, when X is a PHS for a lie group G, the graph of the action is an isomorphism and we obtain a map $X \times X \to G$: $(x, y) \to (a(x, y))$ leading to a first order system of finite Lie equations $y_x = \frac{\partial f}{\partial x}(x, a(x, y))$. In order to produce a Lie form, let us first notice that the general solution of the system of infinitesimal equations is $\xi = \lambda^\tau \theta_\tau$ with $\lambda = cst$. Introducing the inverse matrix $(\omega) = (\omega_i^\tau)$ of the *reciprocal distribution* $\alpha = \{\alpha_\tau\}$ made by n vectors commuting with $\{\theta_\tau\}$, we obtain $\lambda = cst \Leftrightarrow [\xi, \alpha] = 0 \Leftrightarrow \mathcal{L}(\xi)\omega = 0$.

Example 3.2. *(Affine and projective structures of the real line) In Example 2.1 with $n = 1$, the special Lie equations are $\Phi(y_2) \equiv y_{xx}/y_x = 0 \Rightarrow \partial_{xx}\xi = 0$ with $q = 2$ and we let the reader check as an exercise that the general Lie equations are:*

$$\frac{y_{xx}}{y_x} + \omega(y)y_x = \omega(x) \Rightarrow \partial_{xx}\xi + \omega(x)\partial_x\xi + \xi\partial_x\omega(x) = 0$$

with no IC. The special section is $\omega(x) = 0$.
We could study in the same way the group of projective transformations of the real line $y = (ax + b)/(cx + d)$ and get with more work the general lie equations:

$$\frac{y_{xxx}}{y_x} - \frac{3}{2}(\frac{y_{xx}}{y_x})^2 + \omega(y)y_x^2 = \omega(x) \Rightarrow \partial_{xxx}\xi + 2\omega(x)\partial_x\xi + \xi\partial_x\omega(x) = 0$$

There is an isomorphism $J_1(\mathcal{F}_{aff}) \simeq \mathcal{F}_{aff} \times_X \mathcal{F}_{proj} : j_1(\omega) \to (\omega, \gamma = \partial_x\omega - (1/2)\omega^2)$.

Example 3.3. $n = 2, q = 1, \Gamma = \{y^1 = f(x^1), y^2 = x^2/(\partial f(x^1)/\partial x^1)\}$ *where f is an arbitrary invertible map. The involutive Lie form is:*

$$\Phi^1(y_1) \equiv y^2 y_1^1 = x^2,$$
$$\Phi^2(y_1) \equiv y^2 y_2^1 = 0,$$
$$\Phi^3(y_1) \equiv \frac{\partial(y^1, y^2)}{\partial(x^1, x^2)} \equiv y_1^1 y_2^2 - y_2^1 y_1^2 = 1$$

We obtain $\mathcal{F} = T^* \times_X \wedge^2 T^*$ and $\omega = (\alpha, \beta)$ where α is a 1-form and β is a 2-form with special section $\omega = (x^2 dx^1, dx^1 \wedge dx^2)$. It follows that $d\alpha/\beta$ is a well defined scalar because $\beta \neq 0$. The Vessiot structure equation is $d\alpha = c\beta$ with a single structure constant c which cannot have anything to do with a Lie algebra. Considering the other section $\bar{\omega} = (dx^1, dx^1 \wedge dx^2)$, we get $\bar{c} = 0$. As $c = -1$ and thus $\bar{c} \neq c$, the equivalence problem $j_1(f)^{-1}(\omega) = \bar{\omega}$ cannot even be solved formally.

Example 3.4. *(Symplectic structure) With $n = 2p, q = 1$ and $\mathcal{F} = \wedge^2 T^*$, let ω be a closed 2-form of maximum rank, that is $d\omega = 0, det(\omega) \neq 0$. The equivalence problem is nothing else than the Darboux problem in analytical mechanics giving the possibility to write locally $\omega = \sum dp \wedge dq$ by using canonical conjugate coordinates $(q, p) = (position, momentum)$.*

Example 3.5. *(Contact structure) With $n = 3, q = 1, w = dx^1 - x^3 dx^2 \Rightarrow w \wedge dw = dx^1 \wedge dx^2 \wedge dx^3$, let us consider $\Gamma = \{f \in aut(X)|j_1(f)^{-1}(w) = \rho w\}$. This is not a Lie form but we get:*

$$j_1(f)^{-1}(dw) = dj_1(f)^{-1}(w) = \rho dw + d\rho \wedge w \Rightarrow j_1(f)^{-1}(w \wedge dw) = \rho^2(w \wedge dw)$$

The corresponding geometric object is thus made by a 1-form density $\omega = (\omega_1, \omega_2, \omega_3)$ that transforms like a 1-form up to the division by the square root of the Jacobian determinant. The unusual general Medolaghi form is:

$$\Omega_i \equiv \omega_r(x)\partial_i\xi^r - (1/2)\omega_i(x)\partial_r\xi^r + \xi^r\partial_r\omega_i(x) = 0$$

In a symbolic way $\omega \wedge d\omega$ is now a scalar and the only Vessiot structure equation is:

$$\omega_1(\partial_2\omega_3 - \partial_3\omega_2) + \omega_2(\partial_3\omega_1 - \partial_1\omega_3) + \omega_3(\partial_1\omega_2 - \partial_2\omega_1) = c$$

For the special section $\omega = (1, -x^3, 0)$ we have $c = 1$. If we choose $\bar{\omega} = (1, 0, 0)$ we may define $\bar{\Gamma}$ by the system $y_2^1 = 0, y_3^1 = 0, y_2^2 y_3^3 - y_3^2 y_2^3 = y_1^1$ but now $\bar{c} = 0$ and the equivalence problem $j_1(f)^{-1}(\omega) = \bar{\omega}$ cannot even be solved formally. These results can be extended to an arbitrary odd dimension with much more work ([24], p 684).

Example 3.6. *(Screw and complex structures) ($n = 2, q = 1$) In 1878 Clifford introduced abstract numbers of the form $x^1 + \epsilon x^2$ with $\epsilon^2 = 0$ in order to study helicoidal movements in the mechanics of rigid bodies. We may try to define functions of these numbers for which a derivative may have a meaning. Thus, if $f(x^1 + \epsilon x^2) = f^1(x^1, x^2) + \epsilon f^2(x^1, x^2)$, then we should get:*

$$df = (A + \epsilon B)(dx^1 + \epsilon dx^2) = A dx^1 + \epsilon(B dx^1 + A dx^2) = df^1 + \epsilon df^2$$

Accordingly, we have to look for transformations $y^1 = f^1(x^1, x^2), y^2 = f^2(x^1, x^2)$ satisfying the first order involutive system of finite Lie equations $y_2^1 = 0, \quad y_2^2 - y_1^1 = 0$ with no CC. As we have an algebraic Lie pseudogroup, a tricky computation ([24], p 467) allows to prove that Γ is made by the transformations preserving a mixed tensor with square equal to zero as follows:

$$\begin{pmatrix} y_1^1 & y_2^1 \\ y_1^2 & y_2^2 \end{pmatrix}^{-1} \begin{pmatrix} 0 & 0 \\ 1 & 0 \end{pmatrix} \begin{pmatrix} y_1^1 & y_2^1 \\ y_1^2 & y_2^2 \end{pmatrix} = \begin{pmatrix} 0 & 0 \\ 1 & 0 \end{pmatrix}$$

We get the Lie form $\Phi^1 \equiv y_2^1/y_1^1 = 0, \Phi^2 \equiv (y_1^1)^2/(y_1^1 y_2^2 - y_2^1 y_1^2) = 1$ and let the reader exhibit $\mathcal{F}$. Finally, introducing similarly the abstract number i such that $i^2 = -1$, we get the Cauchy-Riemann system $y_2^2 - y_1^1 = 0, \quad y_2^1 + y_1^2 = 0$ with no CC defining complex analytic transformations and the corresponding geometric object or complex structure is a mixed tensor with square equal to minus the 2×2 identity matrix as we have now:

$$\begin{pmatrix} y_1^1 & y_2^1 \\ y_1^2 & y_2^2 \end{pmatrix}^{-1} \begin{pmatrix} 0 & -1 \\ 1 & 0 \end{pmatrix} \begin{pmatrix} y_1^1 & y_2^1 \\ y_1^2 & y_2^2 \end{pmatrix} = \begin{pmatrix} 0 & -1 \\ 1 & 0 \end{pmatrix}$$

Example 3.7. *(Riemann structure) If ω is a section of $\mathcal{F} = S_2 T^*$ with $\det(\omega) \neq 0$ we get:*

Lie form $\Phi_{ij}(y_1) \equiv \omega_{kl}(y)y_i^k y_j^l = \omega_{ij}(x)$

Medolaghi form $\Omega_{ij} \equiv (\mathcal{L}(\xi)\omega)_{ij} \equiv \omega_{rj}(x)\partial_i\xi^r + \omega_{ir}(x)\partial_j\xi^r + \xi^r\partial_r\omega_{ij}(x) = 0$

also called Killing system for historical reasons. A special section could be the Euclidean metric when $n = 1, 2, 3$ as in elasticity theory or the Minkowski metric when $n = 4$ as in special relativity. The main problem is that this system is not involutive unless we prolong the system to order two by differentiating once the equations. For such a purpose, introducing $\omega^{-1} = (\omega^{ij})$ as usual, we may define:

Christoffel symbols $\gamma_{ij}^k(x) = \frac{1}{2}\omega^{kr}(x)(\partial_i\omega_{rj}(x) + \partial_j\omega_{ri}(x) - \partial_r\omega_{ij}(x)) = \gamma_{ji}^k(x)$

This is a new geometric object of order 2 allowing to obtain, as in Example 3.2, an isomorphism $j_1(\omega) \simeq (\omega, \gamma)$ and the second order equations with $f_1^{-1} = g_1$:

Lie form
$$g_l^k(y_{ij}^l + \gamma_{rs}^l(y)y_i^r y_j^s) = \gamma_{ij}^k(x)$$

Medolaghi form $\quad \Gamma_{ij}^k \equiv (\mathcal{L}(\xi)\gamma)_{ij}^k \equiv \partial_{ij}\xi^k + \gamma_{rj}^k(x)\partial_i\xi^r + \gamma_{ir}^k(x)\partial_j\xi^r - \gamma_{ij}^r(x)\partial_r\xi^k + \xi^r\partial_r\gamma_{ij}^k(x) = 0$

where (Γ_{ij}^k) is a section of $S_2T^* \otimes T$. Surprisingly, the following expression:

Riemann tensor
$$\rho_{lij}^k(x) \equiv \partial_i\gamma_{lj}^k(x) - \partial_j\gamma_{li}^k(x) + \gamma_{lj}^r(x)\gamma_{ri}^k(x) - \gamma_{li}^r(x)\gamma_{rj}^k(x)$$

is still a first order geometric object and even a tensor as a section of $\wedge^2 T^* \otimes T^* \otimes T$ satisfying the purely algebraic relations :

$$\rho_{lij}^k + \rho_{ijl}^k + \rho_{jli}^k = 0, \qquad \omega_{rl}\rho_{kij}^l + \omega_{kr}\rho_{lij}^r = 0 \Rightarrow \rho_{klij} = \omega_{kr}\rho_{lij}^r = \rho_{ijkl}.$$

Accordingly, the IC must express that the new first order equations $(\mathcal{L}(\xi)\rho)_{lij}^k = 0$ are only linear combinations of the previous ones and we get the Vessiot structure equations:

$$\rho_{lij}^k(x) = c(\delta_i^k\omega_{lj}(x) - \delta_j^k\omega_{li}(x))$$

describing the constant Riemannian curvature condition of Eisenhart [10]. Finally, as we have $\rho_{rij}^r(x) = \partial_i\gamma_{rj}^r(x) - \partial_j\gamma_{ri}^r(x) = 0$, we can only introduce the Ricci tensor $\rho_{ij}(x) = \rho_{irj}^r(x) = \rho_{ji}(x)$ by contracting indices and the scalar curvature $\rho(x) = \omega^{ij}(x)\rho_{ij}(x)$ in order to obtain $\rho(x) = n(n-1)c$. It remains to obtain all these results in a purely formal way, for example to prove that the number of components of the Riemann tensor is equal to $n^2(n^2 - 1)/12$ without dealing with indices.

Remark 3.2. *Comparing the various Vessiot structure equations containing structure constants, we discover at once that the many c appearing in the MC equations are absolutely on equal footing with the only c appearing in the other examples. As their factors are either constant, linear or quadratic, any identification of the quadratic terms appearing in the Riemann tensor with the quadratic terms appearing in the MC equations is definitively not correct or, in an equivalent but more abrupt way, the Cartan structure equations have nothing to do with the Vessiot structure equations. As we shall see, most of mathematical physics today is based on such a confusion.*

Remark 3.3. *Let us consider again Example 3.2 with $\partial_{xx}f(x)/\partial_x f(x) = \bar{\omega}(x)$ and introduce a variation $\eta(f(x)) = \delta f(x)$ as in analytical or continuum mechanics. We get similarly $\delta\partial_x f = \partial_x\delta f = \frac{\partial\eta}{\partial y}\partial_x f$ and so on, a result leading to $\delta\bar{\omega}(x) = \partial_x f\mathcal{L}(\eta)\omega(f(x))$ where the Lie derivative involved is computed over the target. Let us now pass from the target to the source by introducing $\eta = \xi\partial_x f \Rightarrow \frac{\partial\eta}{\partial y}\partial_x f = \partial_x\xi\partial_x f + \xi\partial_{xx}f$ and so on, a result leading to the particularly simple variation $\delta\bar{\omega} = \mathcal{L}(\xi)\bar{\omega}$ over the soure. As another example of this general variational procedure, let us compare with the similar variations on which classical finite elasticity theory is based. Starting now with $\omega_{kl}(f(x))\partial_i f^k(x)\partial_j f^l(x) = \bar{\omega}_{ij}(x)$, where ω is the Euclidean metric, we obtain $(\delta\bar{\omega})_{ij}(x) = \partial_i f^k(x)\partial_j f^l(x)(\mathcal{L}(\eta)\omega)_{kl}(f(x))$ where the Lie derivative involved is computed over the target. Passing now from the target to the source as before, we find the particularly simple variation $\delta\bar{\omega} = \mathcal{L}(\xi)\bar{\omega}$ over the source. For "small" deformations, source and target are of course identified but it is not true that the infinitesimal deformation tensor is in general the limit of the finite deformation tensor (for a counterexample, see [25], p 70).*

Introducing a copy Y of X in the general framework, $(f, \delta f)$ must be considered as a section of $V(X \times Y) = (X \times Y) \times_Y T(Y) = X \times T(Y)$ over X. When f is invertible (care), then we may consider the map $f : X \to Y : (x) \to (y = f(x))$ and define $\xi \in T$ by $\eta = T(f)(\xi)$ or rather $\eta = j_1(f)(\xi)$ in the language of geometric object, as a way to identify $f^{-1}(V(X \times Y))$ with $T = T(X)$. When $f = id$, this identification is canonical by considering vertical vectors along the diagonal $\Delta = \{(x, y) \in X \times Y | y = x\}$ and we get $\delta \omega = \Omega \in F_0 = \omega^{-1}(V(\mathcal{F}))$. We point out that the above *vertical procedure* is a nice tool for studying nonlinear systems ([26], III, C and [27], III, 2).

4. Janet versus Spencer : The linear sequences

Let $\mu = (\mu_1, ..., \mu_n)$ be a multi-index with *length* $|\mu| = \mu_1 + ... + \mu_n$, *class i* if $\mu_1 = ... = \mu_{i-1} = 0, \mu_i \neq 0$ and $\mu + 1_i = (\mu_1, ..., \mu_{i-1}, \mu_i + 1, \mu_{i+1}, ..., \mu_n)$. We set $y_q = \{y_\mu^k | 1 \leq k \leq m, 0 \leq |\mu| \leq q\}$ with $y_\mu^k = y^k$ when $|\mu| = 0$. If E is a vector bundle over X with local coordinates (x^i, y^k) for $i = 1, ..., n$ and $k = 1, ..., m$, we denote by $J_q(E)$ the *q-jet bundle* of E with local coordinates simply denoted by (x, y_q) and *sections* $f_q : (x) \to (x, f^k(x), f_i^k(x), f_{ij}^k(x), ...)$ transforming like the section $j_q(f) : (x) \to (x, f^k(x), \partial_i f^k(x), \partial_{ij} f^k(x), ...)$ when f is an arbitrary section of E. Then both $f_q \in J_q(E)$ and $j_q(f) \in J_q(E)$ are over $f \in E$ and the *Spencer operator* just allows to distinguish them by introducing a kind of *"difference"* through the operator $D : J_{q+1}(E) \to T^* \otimes J_q(E) : f_{q+1} \to j_1(f_q) - f_{q+1}$ with local components $(\partial_i f^k(x) - f_i^k(x), \partial_i f_j^k(x) - f_{ij}^k(x), ...)$ and more generally $(D f_{q+1})_{\mu,i}^k(x) = \partial_i f_\mu^k(x) - f_{\mu+1_i}^k(x)$. In a symbolic way, *when changes of coordinates are not involved*, it is sometimes useful to write down the components of D in the form $d_i = \partial_i - \delta_i$ and the restriction of D to the kernel $S_{q+1}T^* \otimes E$ of the canonical projection $\pi_q^{q+1} : J_{q+1}(E) \to J_q(E)$ is *minus* the *Spencer map* $\delta = dx^i \wedge \delta_i : S_{q+1}T^* \otimes E \to T^* \otimes S_q T^* \otimes E$. The kernel of D is made by sections such that $f_{q+1} = j_1(f_q) = j_2(f_{q-1}) = ... = j_{q+1}(f)$. Finally, if $R_q \subset J_q(E)$ is a *system* of order q on E locally defined by linear equations $\Phi^\tau(x, y_q) \equiv a_k^{\tau\mu}(x) y_\mu^k = 0$ and local coordinates (x, z) for the parametric jets up to order q, the *r-prolongation* $R_{q+r} = \rho_r(R_q) = J_r(R_q) \cap J_{q+r}(E) \subset J_r(J_q(E))$ is locally defined when $r = 1$ by the linear equations $\Phi^\tau(x, y_q) = 0, d_i \Phi^\tau(x, y_{q+1}) \equiv a_k^{\tau\mu}(x) y_{\mu+1_i}^k + \partial_i a_k^{\tau\mu}(x) y_\mu^k = 0$ and has *symbol* $g_{q+r} = R_{q+r} \cap S_{q+r}T^* \otimes E \subset J_{q+r}(E)$ if one looks at the *top order terms*. If $f_{q+1} \in R_{q+1}$ is over $f_q \in R_q$, differentiating the identity $a_k^{\tau\mu}(x) f_\mu^k(x) \equiv 0$ with respect to x^i and substracting the identity $a_k^{\tau\mu}(x) f_{\mu+1_i}^k(x) + \partial_i a_k^{\tau\mu}(x) f_\mu^k(x) \equiv 0$, we obtain the identity $a_k^{\tau\mu}(x)(\partial_i f_\mu^k(x) - f_{\mu+1_i}^k(x)) \equiv 0$ and thus the restriction $D : R_{q+1} \to T^* \otimes R_q$ ([23],[27],[33]).

Definition 4.1. R_q *is said to be formally integrable when the restriction* $\pi_q^{q+1} : R_{q+1} \to R_q$ *is an epimorphism* $\forall r \geq 0$ *or, equivalently, when all the equations of order* $q + r$ *are obtained by* r *prolongations only* $\forall r \geq 0$. *In that case,* $R_{q+1} \subset J_1(R_q)$ *is a canonical equivalent formally integrable first order system on* R_q *with no zero order equations, called the Spencer form.*

Definition 4.2. R_q *is said to be involutive when it is formally integrable and all the sequences* ... $\xrightarrow{\delta}$ $\wedge^s T^* \otimes g_{q+r} \xrightarrow{\delta}$ ... *are exact* $\forall 0 \leq s \leq n, \forall r \geq 0$. *Equivalently, using a linear change of local coordinates if necessary, we may successively solve the maximum number* $\beta_q^n, \beta_q^{n-1}, ..., \beta_q^1$ *of equations with respect to the principal jet coordinates of strict order* q *and class* $n, n - 1, ..., 1$ *in order to introduce* the characters $\alpha_q^i = m \frac{(q+n-i-1)!}{(q-1)!((n-i)!} - \beta_q^i$ for $i = 1, ..., n$ with $\alpha_q^n = \alpha$. Then R_q is involutive if R_{q+1} is obtained by only prolonging the β_q^i equations of class i with respect to $d_1, ..., d_i$ for $i = 1, ..., n$. In that case $\dim(g_{q+1}) = \alpha_q^1 + ... + \alpha_q^n$ and one can exhibit the Hilbert polynomial $\dim(R_{q+r})$ in r with leading term $(\alpha/n!)r^n$ when $\alpha \neq 0$. Such a prolongation procedure allows to compute in a unique way the principal (pri) jets from the parametric (par) other ones. This definition may also be applied to nonlinear systems as well.

We obtain the following theorem generalizing for PD control systems the well known first order Kalman form of OD control systems where the derivatives of the input do not appear ([27], VI,1.14, p 802):

Theorem 4.1. When R_q is involutive, its Spencer form is involutive and can be modified to a reduced Spencer form in such a way that $\beta = \dim(R_q) - \alpha$ equations can be solved with respect to the jet coordinates $z_n^1, ..., z_n^\beta$ while $z_n^{\beta+1}, ..., z_n^{\beta+\alpha}$ do not appear. In this case $z^{\beta+1}, ..., z^{\beta+\alpha}$ do not appear in the other equations.

When R_q is involutive, the linear differential operator $\mathcal{D} : E \xrightarrow{j_q} J_q(E) \xrightarrow{\Phi} J_q(E)/R_q = F_0$ of order q with space of solutions $\Theta \subset E$ is said to be *involutive* and one has the canonical *linear Janet sequence* ([4], p 144):

$$0 \longrightarrow \Theta \longrightarrow T \xrightarrow{\mathcal{D}} F_0 \xrightarrow{\mathcal{D}_1} F_1 \xrightarrow{\mathcal{D}_2} ... \xrightarrow{\mathcal{D}_n} F_n \longrightarrow 0$$

where each other operator is first order involutive and generates the *compatibility conditions* (CC) of the preceding one. As the Janet sequence can be cut at any place, *the numbering of the Janet bundles has nothing to do with that of the Poincaré sequence*, contrary to what many physicists believe.

Definition 4.3. *The Janet sequence is said to be locally exact at F_r if any local section of F_r killed by $\mathcal{D}_{r+1}$ is the image by $\mathcal{D}_r$ of a local section of F_{r-1}. It is called locally exact if it is locally exact at each F_r for $0 \leq r \leq n$. The Poincaré sequence is locally exact but counterexamples may exist ([23], p 202).*

Equivalently, we have the involutive *first Spencer operator* $D_1 : C_0 = R_q \xrightarrow{j_1} J_1(R_q) \to J_1(R_q)/R_{q+1} \simeq T^* \otimes R_q/\delta(g_{q+1}) = C_1$ of order one induced by $D : R_{q+1} \to T^* \otimes R_q$. Introducing the *Spencer bundles* $C_r = \wedge^r T^* \otimes R_q/\delta(\wedge^{r-1}T^* \otimes g_{q+1})$, the first order involutive $(r + 1)$-*Spencer operator* $D_{r+1} : C_r \to C_{r+1}$ is induced by $D : \wedge^r T^* \otimes R_{q+1} \to \wedge^{r+1}T^* \otimes R_q : \alpha \otimes \xi_{q+1} \to d\alpha \otimes \xi_q + (-1)^r \alpha \wedge D\xi_{q+1}$ and we obtain the canonical *linear Spencer sequence* ([4], p 150):

$$0 \longrightarrow \Theta \xrightarrow{j_q} C_0 \xrightarrow{D_1} C_1 \xrightarrow{D_2} C_2 \xrightarrow{D_3} ... \xrightarrow{D_n} C_n \longrightarrow 0$$

as the Janet sequence for the first order involutive system $R_{q+1} \subset J_1(R_q)$.

The Janet sequence and the Spencer sequence are connected by the following *crucial commutative diagram* (1) where the Spencer sequence is induced by the locally exact central horizontal sequence which is at the same time the Janet sequence for j_q and the Spencer sequence for $J_{q+1}(E) \subset J_1(J_q(E))$ ([25], p 152):

$$SPENCER \quad SEQUENCE$$

$$
\begin{array}{ccccccc}
0 & 0 & 0 & & 0 \\
\downarrow & \downarrow & \downarrow & & \downarrow \\
0 \xrightarrow{\quad} \Theta \xrightarrow{j_q} & C_0 \xrightarrow{D_1} & C_1 \xrightarrow{D_2} & C_2 \xrightarrow{D_3} \ldots \xrightarrow{D_n} & C_n \xrightarrow{\quad} 0 \\
\downarrow & \downarrow & \downarrow & & \downarrow \\
0 \xrightarrow{\quad} E \xrightarrow{j_q} & C_0(E) \xrightarrow{D_1} & C_1(E) \xrightarrow{D_2} & C_2(E) \xrightarrow{D_3} \ldots \xrightarrow{D_n} & C_n(E) \xrightarrow{\quad} 0 \\
\| & \downarrow \Phi_0 & \downarrow \Phi_1 & \downarrow \Phi_2 & \downarrow \Phi_n \\
0 \xrightarrow{\quad} \Theta \xrightarrow{\quad} E \xrightarrow{D} & F_0 \xrightarrow{D_1} & F_1 \xrightarrow{D_2} & F_2 \xrightarrow{D_3} \ldots \xrightarrow{D_n} & F_n \xrightarrow{\quad} 0 \\
\downarrow & \downarrow & \downarrow & & \downarrow \\
0 & 0 & 0 & & 0
\end{array}
$$

$$JANET \quad SEQUENCE$$

In this diagram, *only depending on the left commutative square* $\mathcal{D} = \Phi \circ j_q$, *the epimorphisms*

$\Phi_r : C_r(E) \to F_r$ for $0 \leq r \leq n$ are successively induced by the canonical projection $\Phi = \Phi_0 :$ $C_0(E) = J_q(E) \to J_q(E)/R_q = F_0$.

Example 4.1. *(Screw structure): The system $R_1 \subset J_1(T)$ defined by $\zeta_2^1 = 0, \zeta_2^2 - \zeta_1^1 = 0$ is involutive with $par(R_2) = \{\xi^1, \xi^2, \zeta_1^1, \zeta_1^2, \zeta_{11}^1, \zeta_{11}^2\}$. The Spencer operator is not involutive as it is not even formally integrable because $\partial_2 \zeta_1^2 - \zeta_{11}^1 = 0, \partial_1 \zeta_1^2 - \zeta_{11}^2 = 0 \Rightarrow \partial_1 \zeta_{11}^1 - \partial_2 \zeta_{11}^2 = 0$. We obtain $dim(F_0) = 2, dim(C_0(T)) = 6 \Rightarrow dim(C_0) = dim(R_1) = 4, dim(F_1) = 0 \Rightarrow dim(C_1(T)) = dim(C_1) = 6, dim(C_2(T)) = dim(C_2) = 2$ and it is not evident at all that the first order involutive operator $D_1 : C_0 \to C_1$ is defined by the 6 PD equations for 4 unknowns:*

$$\partial_2 \zeta^1 = 0, \partial_2 \zeta^2 - \zeta_1^1 = 0, \partial_2 \zeta_1^1 = 0, \partial_2 \zeta_1^2 - \partial_1 \zeta_1^1 = 0, \partial_1 \zeta^1 - \zeta_1^1 = 0, \partial_1 \zeta^2 - \zeta_1^2 = 0$$

The case of a complex structure is similar and left to the reader.

5. Differential modules and inverse systems

An important but difficult problem in engineering physics is to study how the formal properties of a system of order q with n independent variables and m unknowns depend on the parameters involved in that system. This is particularly clear in classical control theory where the systems are classified into two categories, namely the "controllable" ones and the "uncontrollable" ones ([14],[27]). In order to understand the problem studied by Macaulay in [M], that is roughly to determine the minimum number of solutions of a system that must be known in order to determine all the others by using derivatives and linear combinations with constant coefficients in a field k, let us start with the following motivating example:

Example 5.1. *When $n = 1, m = 1, q = 3$, using a sub-index x for the derivatives with $d_{xy} = y_x$ and so on, the general solution of $y_{xxx} - y_x = 0$ is $y = ae^x + be^{-x} + c1$ with a, b, c constants and the derivative of e^x is e^x, the derivative of e^{-x} is $-e^{-x}$ and the derivative of 1 is 0. Hence we could believe that we need a basis $\{1, e^x, e^{-x}\}$ with three generators for obtaining all the solutions through derivatives. Also, when $n = 1, m = 2, k = \mathbb{R}$ and a is a constant real parameter, the OD system $y_{xx}^1 - ay^1 = 0, y_x^2 = 0$ needs two generators $\{(x, 0), (0, 1)\}$ when $a = 0$ with the only d_x killing both y_x^1 and y_2 but only one generator when $a \neq 0$, namely $\{(ch(x), 1)\}$ when $a = 1$. Indeed, setting $y = y^1 - y^2$ brings $y^1 = y_{xx}, y^2 = y_{xx} - y$ and an equivalent system defined by the* *single OD equation* $y_{xxx} - y_x = 0$ *for the only y. Introducing the corresponding polynomial ideal* $(\chi^3 - \chi) = (\chi) \cap (\chi - 1) \cap (\chi + 1)$, *we check that* d_x *kills* $y_{xx} - y$, $d_x - 1$ *kills* $y_{xx} + y_x$ *and* $d_x + 1$ *kills* $y_{xx} - y_x$, *a result leading, as we shall see, to the only generator* $\{ch(x) - 1\}$.

More precisely, if K is a differential field containing $\mathbb{Q}$ with n commuting *derivations* ∂_i, that is to say $\partial_i(a + b) = \partial_i a + \partial_i b$ and $\partial_i(ab) = (\partial_i a)b + a\partial_i b, \forall a, b \in K$ for $i = 1, ..., n$, we denote by k a subfield of constants. Let us introduce m *differential indeterminates* y^k for $k = 1, ..., m$ and n commuting *formal derivatives* d_i with $d_i y^k_\mu = y^k_{\mu + 1_i}$. We introduce the non-commutative *ring of differential operators* $D = K[d_1, ..., d_n] = K[d]$ with $d_i a = ad_i + \partial_i a, \forall a \in K$ in the operator sense and the *differential module* $Dy = Dy^1 + ... + Dy^m$. If $\{\Phi^\tau = a_k^{\tau\mu} y^k_\mu\}$ is a finite number of elements in Dy indexed by τ, we may introduce the *differential module of equations* $I = D\Phi \subset Dy$ and the finitely generated *residual differential module* $M = Dy/I$.

In the algebraic framework considered, *only two possible formal constructions can be obtained from* M when $D = K[d]$, namely $hom_D(M, D)$ and $M^* = hom_K(M, K)$ ([3],[27],[32]).

Theorem 5.1. $hom_D(M, D)$ *is a right differential module that can be converted to a left differential module by introducing the right differential module structure of* $\wedge^n T^*$. *As a differential geometric counterpart, we get the formal adjoint of* $\mathcal{D}$, *namely* $ad(\mathcal{D}) : \wedge^n T^* \otimes F^* \to \wedge^n T^* \otimes E^*$ *usually constructed through an integration by parts and where* E^* *is obtained from E by inverting the local transition matrices, the simplest example being the way* T^* *is obtained from T.*

Remark 5.1. *Such a result explains why dual objects in physics and engineering are no longer tensors but tensor densities, with no reference to any variational calculus. For example the EM potential is a section of* T^* *and the EM field is a section of* $\wedge^2 T^*$ *while the EM induction is a section of* $\wedge^4 T^* \otimes \wedge^2 T \simeq \wedge^2 T^*$ *and the EM current is a section of* $\wedge^4 T^* \otimes T \simeq \wedge^3 T^*$ *when* $n = 4$.

The filtration $D_0 = K \subseteq D_1 = K \oplus T \subseteq ... \subseteq D_q \subseteq ... \subseteq D$ of D by the order of operators induces a filtration/inductive limit $0 \subseteq M_0 \subseteq M_1 \subseteq ... \subseteq M_q \subseteq ... \subseteq M$ and provides by duality *over* K the projective limit $M^* = R \to ... \to R_q \to ... \to R_1 \to R_0 \to 0$ of formally integrable systems. As D is generated by K and $T = D_1/D_0$, we can define for any $f \in M^*$:

$$(af)(m) = af(m) = f(am), (\xi f)(m) = \xi f(m) - f(\xi m), \forall a \in K, \forall \xi = a^i d_i \in T, \forall m \in M$$

and check $d_i a = ad_i + \partial_i a, \xi\eta - \eta\xi = [\xi, \eta]$ in the operator sense by introducing the standard bracket of vector fields on T. Finally we get $(d_i f)^k_\mu = (d_i f)(y^k_\mu) = \partial_i f^k_\mu - f^k_{\mu + 1_i}$ in a coherent way.

Theorem 5.2. $R = M^*$ *has a structure of differential module induced by the Spencer operator.*

Remark 5.2. *When* $m = 1$ *and* $D = k[d]$ *is a commutative ring isomorphic to the polynomial ring* $A = k[\chi]$ *for the indeterminates* $\chi_1, ..., \chi_n$, *this result exactly describes the inverse system of Macaulay with* $-d_i = \delta_i$ *([M], §59,60).*

Definition 5.1. *A simple module is a module having no other proper submodule than 0. A semi-simple module is a direct sum of simple modules. When A is a commutative integral domain and M a finitely generated module over A, the socle of M is the largest semi-simple submodule of M, that is* $soc(M) = \oplus soc_{\mathfrak{m}}(M)$ *where* $soc_{\mathfrak{m}}(M)$ *is the direct sum of all the isotypical simple submodules of M isomorphic to* $A/\mathfrak{m}$ *for* $\mathfrak{m} \in max(A)$ *the set of maximal proper ideals of A. The radical of a module is the intersection of all its maximum proper submodules. The quotient of a module by its radical is called the top and is a semi-simple module ([3]).*

The "*secret* " of Macaulay is expressed by the next theorem:

Theorem 5.3. *Instead of using the socle of M over A, one may use duality over k in order to deal with the short exact sequence* $0 \to rad(R) \to R \to top(R) \to 0$ *where* $top(R)$ *is the dual of* $soc(M)$.

However, Nakayama's lemma ([3],[19],[32]) cannot be used in general unless R is finitely generated over k and thus over D. The main idea of Macaulay has been to overcome this difficulty by dealing only with *unmixed* ideals when $m = 1$. As a generalization, one can state ([27]):

Definition 5.2. *One has the purity filtration* $0 = t_n(M) \subseteq ... \subseteq t_0(M) = t(M) \subseteq M$ *where any involutive system of order p defining Dm is such that* $\alpha_p^{n-r} = 0, ..., \alpha_p^n = 0$ *when* $m \in t_r(M)$ *and M is said to be r-pure if* $t_r(M) = 0, t_{r-1}(M) = M$. *With* $t(M) = \{m \in M \mid \exists 0 \neq a \in A, am = 0\}$ *we say that M is a 0-pure or torsion-free module if* $t(M) = 0$ *and a torsion module if* $t(M) = M$.

Example 5.2. *With* $n = 2, q = 2$, *let us consider the involutive system* $y_{(0,2)} \equiv y_{22} = 0, y_{(1,1)} \equiv y_{12} = 0$. *Then* $z' = y_1$ *satisfies* $z'_2 = 0$ *while* $z'' = y_2$ *satisfies* $z''_2 = 0, z''_1 = 0$ *and we have the filtration* $0 = t_2(M) \subset t_1(M) \subset t_0(M) = t(M) = M$ *with* $z'' \in t_1(M), z' \in t_0(M)$ *but* $z' \notin t_1(M)$. *This classification of observables has never been applied to engineering systems like the ones to be found in magnetohydrodynamics (MHD) because the mathematics involved are not known.*

Remark 5.3. *A standard result in commutative algebra allows to embed any torsion-free module into a free module ([32]). Such a property provides the possibility to parametrize the solution space of the corresponding system of OD/PD equations by a finite number of potential like arbitrary functions. For this, in order to test the possibility to parametrize a given operator* $\mathcal{D}_1$, *one may construct the adjoint operator* $ad(\mathcal{D}_1)$ *and look for generating CC in the form of an operator* $ad(\mathcal{D})$. *As* $ad(\mathcal{D}) \circ ad(\mathcal{D}_1) = ad(\mathcal{D}_1 \circ \mathcal{D}) = 0 \Rightarrow \mathcal{D}_1 \circ \mathcal{D} = 0$, *it only remains to check that the CC of* $\mathcal{D}$ *are generated by* $\mathcal{D}_1$. *When* $n = 1$ *this result amounts to Kalman test and the fact that a classical OD control system is controllable if and only if it is parametrizable, a result showing that controllability is an intrinsic structural property of a control system, not depending on the choice of inputs and outputs contrary to a well established engineering tradition ([14],[27]). When* $n = 2$, *the formal adjoint of the only CC for the deformation tensor has been used in the Introduction in order to parametrize the stress equation by means of the Airy function. This result is also valid for the non-commutative ring* $D = K[d]$.

Example 5.3. *With* $K = \mathbb{Q}(x^1, x^2, x^3)$, *infinitesimal contact transformations are defined by the system* $\partial_2\xi^1 - x^3\partial_2\xi^2 + x^3\partial_1\xi^1 - (x^3)^2\partial_1\xi^2 - \xi^3 = 0, \quad \partial_3\xi^1 - x^3\partial_3\xi^2 = 0$. *Multiplying by test functions* (λ^1, λ^2) *and integrating by parts, we obtain the adjoint operator (up to sign):*

$$\partial_2\lambda^1 + x^3\partial_1\lambda^1 + \partial_3\lambda^2 = \mu^1, \quad -x^3\partial_2\lambda^1 - (x^3)^2\partial_1\lambda^1 - x^3\partial_3\lambda^2 - \lambda^2 = \mu^2, \quad \lambda^1 = \mu^3$$

It follows that $\lambda^1 = \mu^3, \lambda^2 = -\mu^2 - x^3\mu^1 \Rightarrow \partial_2\mu^3 + x^3\partial_1\mu^3 - \partial_3\mu^2 - x^3\partial_3\mu^1 - 2\mu^1 = 0$. *Multiplying again by a test function* ϕ, *we discover the parametrization* $\xi^1 = x^3\partial_3\phi - \phi, \xi^2 = \partial_3\phi, \xi^3 = -\partial_2\phi - x^3\partial_1\phi$ *which is not evident at first sight.*

When M is r-pure, Theorem 4.3 provides the exact sequence $0 \to M \to k(\chi_1, ..., \chi_{n-r}) \otimes M$, also discovered by Macaulay ([M], §77, 82), and one obtains the following key result for studying the *identifiability* of OD/PD control systems (see *localization* in ([19],[27],32[29],[30],[32]).

Theorem 5.4. *When M is n-pure, one may use the chinese remainder theorem ([19], p 41) in order to prove that the minimum number of generators of R is equal to the maximum number of isotypical components that can be found among the various components of soc(M) or top(R). When M is r-pure but $r \leq n - 1$, the minimum number of generators of R is smaller or equal to the smallest non-zero character.*

6. Janet versus Spencer : The nonlinear sequences

Nonlinear operators do not in general admit CC as can be seen by considering the involutive example $y_{22} - \frac{1}{3}(y_{11})^3 = u, y_{12} - \frac{1}{2}(y_{11})^2 = v$ with $m = 1, n = 2, q = 2$, contrary to what happens in the study of Lie pseudogroups. However, the kernel of a linear operator $\mathcal{D} : E \to F$ is always taken with respet to the zero section of F, while it must be taken with respect to a prescribed section by a *double arrow* for a nonlinear operator. Keeping in mind the linear Janet sequence and the examples of Vessiot structure equations already presented, one obtains:

Theorem 6.1. *There exists a nonlinear Janet sequence associated with the Lie form of an involutive system of finite Lie equations:*

$$0 \to \Gamma \to aut(X) \underset{\omega \circ \alpha}{\overset{\Phi_\omega \circ j_q}{\rightrightarrows}} \mathcal{F} \underset{0}{\overset{I \circ j_1}{\rightrightarrows}} \mathcal{F}_1$$

where the kernel of the first operator $f \to \Phi_\omega \circ j_q(f) = \Phi_\omega(j_q(f)) = j_q(f)^{-1}(\omega)$ is taken with respect to the section ω of $\mathcal{F}$ while the kernel of the second operator $\omega \to I(j_1(\omega)) \equiv A(\omega)\partial_x\omega + B(\omega)$ is taken with respect to the zero section of the vector bundle $\mathcal{F}_1$ over $\mathcal{F}$.

Corollary 6.1. *By linearization at the identity, one obtains the involutive Lie operator $\mathcal{D} = \mathcal{D}_\omega$: $T \to F_0 : \xi \to \mathcal{L}(\xi)\omega$ with kernel $\Theta = \{\xi \in T | \mathcal{L}(\xi)\omega = 0\} \subset T$ satisfying $[\Theta, \Theta] \subset \Theta$ and the corresponding linear Janet sequence where $F_0 = \omega^{-1}(V(\mathcal{F}))$ and $F_1 = \omega^{-1}(\mathcal{F}_1)$.*

Now we notice that T is a natural vector bundle of order 1 and $J_q(T)$ is thus a natural vector bundle of order $q + 1$. Looking at the way a vector field and its derivatives are transformed under any $f \in aut(X)$ while replacing $j_q(f)$ by f_q, we obtain:

$$\eta^k(f(x)) = f_r^k(x)\xi^r(x) \Rightarrow \eta_u^k(f(x))f_i^u(x) = f_r^k(x)\xi_i^r(x) + f_{ri}^k(x)\xi^r(x)$$

and so on, a result leading to:

Lemma 6.1. *$J_q(T)$ is associated with $\Pi_{q+1} = \Pi_{q+1}(X, X)$ that is we can obtain a new section $\eta_q = f_{q+1}(\xi_q)$ from any section $\xi_q \in J_q(T)$ and any section $f_{q+1} \in \Pi_{q+1}$ by the formula:*

$$d_\mu\eta^k \equiv \eta_r^k f_\mu^r + ... = f_r^k\xi_\mu^r + ... + f_{\mu+1,r}^k\xi^r, \forall 0 \leq |\mu| \leq q$$

where the left member belongs to $V(\Pi_q)$. Similarly $R_q \subset J_q(T)$ is associated with $\mathcal{R}_{q+1} \subset \Pi_{q+1}$.

In order to construct another nonlinear sequence, we need a few basic definitions on *Lie groupoids* and *Lie algebroids* that will become substitutes for Lie groups and Lie algebras. As in the beginning of section 3, the first idea is to use the chain rule for derivatives $j_q(g \circ f) = j_q(g) \circ j_q(f)$ whenever $f, g \in aut(X)$ can be composed and to replace both $j_q(f)$ and $j_q(g)$ respectively by f_q and g_q in order to obtain the new section $g_q \circ f_q$. This kind of "composition" law can be written in a pointwise symbolic way by introducing another copy

Z of X with local coordinates (z) as follows:

$$\gamma_q : \Pi_q(Y,Z) \times_Y \Pi_q(X,Y) \to \Pi_q(X,Z) : ((y,z,\frac{\partial z}{\partial y},...),(x,y,\frac{\partial y}{\partial x},...) \to (x,z,\frac{\partial z}{\partial y}\frac{\partial y}{\partial x},...)$$

We may also define $j_q(f)^{-1} = j_q(f^{-1})$ and obtain similarly an "inversion" law.

Definition 6.1. *A fibered submanifold $\mathcal{R}_q \subset \Pi_q$ is called a system of finite Lie equations or a Lie groupoid of order q if we have an induced source projection $\alpha_q : \mathcal{R}_q \to X$, target projection $\beta_q : \mathcal{R}_q \to X$, composition $\gamma_q : \mathcal{R}_q \times_X \mathcal{R}_q \to \mathcal{R}_q$, inversion $\iota_q : \mathcal{R}_q \to \mathcal{R}_q$ and identity $id_q : X \to \mathcal{R}_q$. In the sequel we shall only consider transitive Lie groupoids such that the map $(\alpha_q, \beta_q) : \mathcal{R}_q \to X \times X$ is an epimorphism and we shall denote by $\mathcal{R}_q^0 = id^{-1}(\mathcal{R}_q)$ the isotropy Lie group bundle of $\mathcal{R}_q$. Also, one can prove that the new system $\rho_r(\mathcal{R}_q) = \mathcal{R}_{q+r}$ obtained by differentiating r times all the defining equations of $\mathcal{R}_q$ is a Lie groupoid of order $q + r$. Finally, one can write down the Lie form and obtain $\mathcal{R}_q = \{f_q \in \Pi_q | f_q^{-1}(\omega) = \omega\}$.*

Now, using the *algebraic bracket* $\{j_{q+1}(\xi), j_{q+1}(\eta)\} = j_q([\xi,\eta]), \forall \xi, \eta \in T$, we may obtain by bilinearity a *differential bracket* on $J_q(T)$ extending the bracket on T:

$$[\xi_q, \eta_q] = \{\xi_{q+1}, \eta_{q+1}\} + i(\xi)D\eta_{q+1} - i(\eta)D\xi_{q+1}, \forall \xi_q, \eta_q \in J_q(T)$$

which does not depend on the respective lifts ξ_{q+1} and η_{q+1} of ξ_q and η_q in $J_{q+1}(T)$. This bracket on sections satisfies the Jacobi identity and we set:

Definition 6.2. *We say that a vector subbundle $R_q \subset J_q(T)$ is a system of infinitesimal Lie equations or a Lie algebroid if $[R_q, R_q] \subset R_q$, that is to say $[\xi_q, \eta_q] \in R_q, \forall \xi_q, \eta_q \in R_q$. The kernel R_q^0 of the projection $\pi_0^q : R_q \to T$ is the isotropy Lie algebra bundle of $\mathcal{R}_q^0$ and $[R_q^0, R_q^0] \subset R_q^0$ does not contain derivatives. Such a definition can be checked by means of computer algebra.*

Proposition 6.1. *There is a nonlinear differential sequence:*

$$0 \longrightarrow aut(X) \xrightarrow{j_{q+1}} \Pi_{q+1}(X,X) \xrightarrow{\bar{D}} T^* \otimes J_q(T) \xrightarrow{\bar{D}'} \wedge^2 T^* \otimes J_{q-1}(T)$$

with $\bar{D}f_{q+1} \equiv f_{q+1}^{-1} \circ j_1(f_q) - id_{q+1} = \chi_q \Rightarrow \bar{D}'\chi_q(\xi,\eta) \equiv D\chi_q(\xi,\eta) - \{\chi_q(\xi), \chi_q(\eta)\} = 0$. Moreover, setting $\chi_0 = A - id \in T^* \otimes T$, this sequence is locally exact if $det(A) \neq 0$.

Proof. There is a canonical inclusion $\Pi_{q+1} \subset J_1(\Pi_q)$ defined by $y_{\mu,i}^k = y_{\mu+1_i}^k$ and the composition $f_{q+1}^{-1} \circ j_1(f_q)$ is a well defined section of $J_1(\Pi_q)$ over the section $f_q^{-1} \circ f_q = id_q$ of Π_q like id_{q+1}. The difference $\chi_q = f_{q+1}^{-1} \circ j_1(f_q) - id_{q+1}$ is thus a section of $T^* \otimes V(\Pi_q)$ over id_q and we have already noticed that $id_q^{-1}(V(\Pi_q)) = J_q(T)$. For $q = 1$ we get with $g_1 = f_1^{-1}$:

$$\chi_{,i}^k = g_l^k \partial_i f^l - \delta_i^k = A_i^k - \delta_i^k, \quad \chi_{j,i}^k = g_l^k(\partial_i f_j^l - A_i^r f_{rj}^l)$$

We also obtain from Lemma 6.1 the useful formula $f_r^k \chi_{\mu,i}^r + ... + f_{\mu+1,}^k \chi_{,i}^r = \partial_i f_\mu^k - f_{\mu+1_i}^k$ allowing to determine χ_q inductively.

We refer to ([26], p 215) for the inductive proof of the local exactness, providing the only formulas that will be used later on and can be checked directly by the reader:

$$\partial_i \chi^k_{l,j} - \partial_j \chi^k_{,i} - \chi^k_{i,j} + \chi^k_{j,i} - (\chi^r_{,i}\chi^k_{r,j} - \chi^r_{,j}\chi^k_{r,i}) = 0$$

$$\partial_i \chi^k_{l,j} - \partial_j \chi^k_{l,i} - \chi^k_{li,j} + \chi^k_{lj,i} - (\chi^r_{,i}\chi^k_{lr,j} + \chi^r_{l,i}\chi^k_{r,j} - \chi^r_{l,j}\chi^k_{r,i} - \chi^r_{,j}\chi^k_{lr,i}) = 0$$

There is no need for double-arrows in this framework as the kernels are taken with respect to the zero section of the vector bundles involved. We finally notice that the main difference with the gauge sequence is that *all the indices range from 1 to n* and that the condition $det(A) \neq 0$ amounts to $\Delta = det(\partial_i f^k) \neq 0$ because $det(f^k_i) \neq 0$ by assumption.   □

Corollary 6.2. *There is a restricted nonlinear differential sequence:*

$$0 \longrightarrow \Gamma \xrightarrow{j_{q+1}} \mathcal{R}_{q+1} \xrightarrow{\bar{D}} T^* \otimes R_q \xrightarrow{\bar{D}'} \wedge^2 T^* \otimes J_{q-1}(T)$$

Definition 6.3. *A splitting of the short exact sequence $0 \to R^0_q \to R_q \xrightarrow{\pi^q_0} T \to 0$ is a map $\chi'_q : T \to R_q$ such that $\pi^q_0 \circ \chi'_q = id_T$ or equivalently a section of $T^* \otimes R_q$ over $id_T \in T^* \otimes T$ and is called a R_q-connection. Its curvature $\kappa'_q \in \wedge^2 T^* \otimes R^0_q$ is defined by $\kappa'_q(\xi, \eta) = [\chi'_q(\xi), \chi'_q(\eta)] - \chi'_q([\xi, \eta])$. We notice that $\chi'_q = -\chi_q$ is a connection with $\bar{D}'\chi'_q = \kappa'_q$ if and only if $A = 0$. In particular $(\delta^k_i, -\gamma^k_{ij})$ is the only existing symmetric connection for the Killing system.*

Remark 6.1. *Rewriting the previous formulas with A instead of χ_0 we get:*

$$\partial_i A^k_j - \partial_j A^k_i - A^r_i \chi^k_{r,j} + A^r_j \chi^k_{r,i} = 0$$

$$\partial_i \chi^k_{l,j} - \partial_j \chi^k_{l,i} - \chi^r_{l,i}\chi^k_{r,j} + \chi^r_{l,j}\chi^k_{r,i} - A^r_i \chi^k_{lr,j} + A^r_j \chi^k_{lr,i} = 0$$

When $q = 1, g_2 = 0$ and though surprising it may look like, we find back exactly all the formulas presented by E. and F. Cosserat in ([C], p 123 and [16]). Even more strikingly, in the case of a Riemann structure, the last two terms disappear but the quadratic terms are left while, in the case of screw and complex structures, the quadratic terms disappear but the last two terms are left.

Corollary 6.3. *When $det(A) \neq 0$ there is a nonlinear stabilized sequence at order q:*

$$0 \longrightarrow aut(X) \xrightarrow{j_q} \Pi_q \xrightarrow{\bar{D}_1} C_1(T) \xrightarrow{\bar{D}_2} C_2(T)$$

called nonlinear Spencer sequence where $\bar{D}_1$ and $\bar{D}_2$ are involutive and its restriction:

$$0 \longrightarrow \Gamma \xrightarrow{j_q} \mathcal{R}_q \xrightarrow{\bar{D}_1} C_1 \xrightarrow{\bar{D}_2} C_2$$

is such that $\bar{D}_1$ and $\bar{D}_2$ are involutive whenever $\mathcal{R}_q$ is involutive.

Proof. : With $|\mu| = q$ we have $\chi^k_{\mu,i} = -g^k_l A^r_i f^l_{\mu+1_r} + terms(order \leq q)$. Setting $\chi^k_{\mu,i} = A^r_i \tau^k_{\mu,r}$, we obtain $\tau^k_{\mu,r} = -g^k_l f^l_{\mu+1_r} + terms(order \leq q)$ and $\bar{D} : \Pi_{q+1} \to T^* \otimes J_q(T)$ restricts to $\bar{D}_1 : \Pi_q \to C_1(T)$.

Finally, setting $A^{-1} = B = id - \tau_0$, we obtain successively:

$$\partial_i \chi^k_{\mu,j} - \partial_j \chi^k_{\mu,i} + terms(\chi_q) - (A^r_i \chi^k_{\mu+1_r,j} - A^r_j \chi^k_{\mu+1_r,i}) = 0$$

$$B^i_r B^j_s (\partial_i \chi^k_{\mu,j} - \partial_j \chi^k_{\mu,i}) + terms(\chi_q) - (\tau^k_{\mu+1_r,s} - \tau^k_{\mu+1_s,r}) = 0$$

We obtain therefore $D\tau_{q+1} + terms(\tau_q) = 0$ and $\bar{D}' : T^* \otimes J_q(T) \to \wedge^2 T^* \otimes J_{q-1}(T)$ restricts to $\bar{D}_2 : C_1(T) \to C_2(T)$.

In the case of Lie groups of transformations, the symbol of the involutive system R_q *must* be $g_q = 0$ providing an isomorphism $\mathcal{R}_{q+1} \simeq \mathcal{R}_q \Rightarrow R_{q+1} \simeq R_q$ and we have therefore $C_r = \wedge^r T^* \otimes R_q$ for $r = 1, ..., n$ in the linear Spencer sequence. □

Remark 6.2. *The passage from χ_q to τ_q is exactly the one done by E. and F. Cosserat in ([C], p 190). However, even if is a good idea to pass from the source to the target, the way they realize it is based on a subtle misunderstanding that we shall correct later on in Proposition 6.3.*

If $f_{q+1}, g_{q+1} \in \Pi_{q+1}$ and $f'_{q+1} = g_{q+1} \circ f_{q+1}$, we get:

$$\bar{D}f'_{q+1} = f^{-1}_{q+1} \circ g^{-1}_{q+1} \circ j_1(g_q) \circ j_1(f_q) - id_{q+1} = f^{-1}_{q+1} \circ \bar{D}g_{q+1} \circ j_1(f_q) + \bar{D}f_{q+1}$$

Definition 6.4. *For any section $f_{q+1} \in \mathcal{R}_{q+1}$, the transformation:*

$$\chi_q \longrightarrow \chi'_q = f^{-1}_{q+1} \circ \chi_q \circ j_1(f_q) + \bar{D}f_{q+1}$$

is called a gauge transformation and exchanges the solutions of the field equations $\bar{D}'\chi_q = 0$.

Introducing the *formal Lie derivative* on $J_q(T)$ by the formulas:

$$L(\xi_{q+1})\eta_q = \{\xi_{q+1}, \eta_{q+1}\} + i(\xi)D\eta_{q+1} = [\xi_q, \eta_q] + i(\eta)D\xi_{q+1}$$

$$(L(j_1(\xi_{q+1}))\chi_q)(\zeta) = L(\xi_{q+1})(\chi_q(\zeta)) - \chi_q([\xi, \zeta])$$

and passing to the limit with $f_{q+1} = id_{q+1} + t\xi_{q+1} + ...$ for $t \to 0$ *over the source*, we get:

Lemma 6.2. *An infinitesimal gauge transformation has the form:*

$$\delta \chi_q = D\xi_{q+1} + L(j_1(\xi_{q+1}))\chi_q$$

Passing again to the limit but now over the target with $\chi_q = \bar{D}f_{q+1}$ and $g_{q+1} = id_{q+1} + t\eta_{q+1} + ...$, we obtain the variation:

$$\delta \chi_q = f^{-1}_{q+1} \circ D\eta_{q+1} \circ j_1(f_q)$$

Proposition 6.2. *The same variation is obtained whenever $\eta_{q+1} = f_{q+2}(\xi_{q+1} + \chi_{q+1}(\xi))$ with $\chi_{q+1} = \bar{D}f_{q+2}$, a transformation which only depends on $j_1(f_{q+1})$ and is invertible if and only if $det(A) \neq 0$.*

Proof. : Choosing $f_{q+1}, g_{q+1}, h_{q+1} \in \mathcal{R}_{q+1}$ such that $g_{q+1} \circ f_{q+1} = f_{q+1} \circ h_{q+1}$ and passing to the limits $g_{q+1} = id_{q+1} + t\eta_{q+1} + \ldots$ and $h_{q+1} = id_{q+1} + t\xi_{q+1} + \ldots$ when $t \to 0$, we obtain the local formula:

$$d_\mu \eta^k = \eta_r^k f_\mu^r + \ldots = \xi^i(\partial_i f_\mu^k - f_{\mu+1_i}^k) + f_{\mu+1_r}^k \xi^r + \ldots + f_r^k \xi_\mu^r$$

and thus $\eta_{q+1} = f_{q+2}(\bar{\xi}_{q+1})$ with $\bar{\xi}_{q+1} = \xi_{q+1} + \chi_{q+1}(\xi)$. This transformation is invertible if and only if $\xi \to \bar{\xi} = \xi + \chi_0(\xi) = A(\xi)$ is an isomorphism of T. $\square$

Example 6.1. *For $q = 1$, we obtain from $\delta\chi_q = D\bar{\xi}_{q+1} - \{\chi_{q+1}, \bar{\xi}_{q+1}\}$:*

$$\delta\chi_{,i}^k = (\partial_i\xi^k - \xi_i^k) + (\xi^r\partial_r\chi_{,i}^k + \chi_{,r}^k\partial_i\xi^r - \chi_{,i}^r\xi_r^k)$$
$$= (\partial_i\bar{\xi}^k - \bar{\xi}_i^k) + (\chi_{,i}^k\bar{\xi}^r - \chi_{,i}^r\bar{\xi}_r^k)$$
$$\delta\chi_{j,i}^k = (\partial_i\xi_j^k - \xi_{ij}^k) + (\xi^r\partial_r\chi_{j,i}^k + \chi_{j,r}^k\partial_i\xi^r + \chi_{r,i}^k\xi_j^r - \chi_{j,i}^r\xi_r^k - \chi_{,i}^r\xi_{jr}^k)$$
$$= (\partial_i\bar{\xi}_j^k - \bar{\xi}_{ij}^k) + (\chi_{rj,i}^k\bar{\xi}^r + \chi_{r,i}^k\bar{\xi}_j^r - \chi_{j,i}^r\bar{\xi}_r^k - \chi_{,i}^r\bar{\xi}_{jr}^k)$$

For the Killing system $R_1 \subset J_1(T)$ with $g_2 = 0$, these variations are exactly the ones that can be found in ([C], (50)+(49), p 124 with a printing mistake corrected on p 128) when replacing a 3×3 skewsymmetric matrix by the corresponding vector. The last unavoidable Proposition is thus essential in order to bring back the nonlinear framework of finite elasticity to the linear framewok of infinitesimal elasticity that only depends on the linear Spencer operator.

For the conformal Killing system $\hat{R}_1 \subset J_1(T)$ (see next section) we obtain:

$$\alpha_i = \chi_{r,i}^r \Rightarrow \delta\alpha_i = (\partial_i\bar{\xi}_r^r - \bar{\xi}_{ri}^r) + (\xi^r\partial_r\alpha_i + \alpha_r\partial_i\xi^r + \chi_{,i}^s\bar{\xi}_{rs}^r)$$

This is exactly the variation obtained by Weyl ([W], (76), p 289) who was assuming implicitly $A = 0$ when setting $\bar{\xi}_r^r = 0 \Leftrightarrow \xi_r^r = -\alpha_i\xi^i$ by introducing a connection. Accordingly, ξ_{ri}^r is the variation of the EM potential itself, that is the δA_i of engineers used in order to exhibit the Maxwell equations from a variational principle ([W], § 26) but the introduction of the Spencer operator is new in this framework.

Finally, chasing in diagram (1) , the Spencer sequence is locally exact at C_1 if and only if the Janet sequence is locally exact at F_0 because the central sequence is locally exact. *The situation is much more complicate in the nonlinear framewok.* Let $\bar{\omega}$ be a section of $\mathcal{F}$ satisfying the same CC as ω, namely $I(j_1(\omega)) = 0$. It follows that we can find a section $f_{q+1} \in \Pi_{q+1}$ such that $f_q^{-1}(\omega) = \bar{\omega} \Rightarrow j_1(f_q^{-1})(j_1(\omega)) = j_1(f_q^{-1}(\omega)) = j_1(\bar{\omega})$ and $f_{q+1}^{-1}(j_1(\omega)) = j_1(\bar{\omega})$. We obtain therefore $j_1(f_q^{-1})(j_1(\omega)) = f_{q+1}^{-1}(j_1(\omega)) \Rightarrow (f_{q+1} \circ j_1(f_q^{-1}))^{-1}(j_1(\omega)) - j_1(\omega) = L(\sigma_q)\omega = 0$ and thus $\sigma_q = \bar{D}f_{q+1}^{-1} \in T^* \otimes R_q$ over the target, even if f_{q+1} may not be a section of $\mathcal{R}_{q+1}$. As σ_q is killed by $\bar{D}'$, we have related cocycles at $\mathcal{F}$ in the Janet sequence *over the source* with cocycles at $T^* \otimes R_q$ or C_1 *over the target*.

Now, if $f_{q+1}, f'_{q+1} \in \Pi_{q+1}$ are such that $f_{q+1}^{-1}(j_1(\omega)) = f'^{-1}_{q+1}(j_1(\omega)) = j_1(\bar{\omega})$, it follows that $(f'_{q+1} \circ f_{q+1}^{-1})(j_1(\omega)) = j_1(\omega) \Rightarrow \exists g_{q+1} \in \mathcal{R}_{q+1}$ such that $f'_{q+1} = g_{q+1} \circ f_{q+1}$ and the new $\sigma'_q = \bar{D}f'^{-1}_{q+1}$ differs from the initial $\sigma_q = \bar{D}f_{q+1}^{-1}$ by a gauge transformation.

Conversely, let $f_{q+1}, f'_{q+1} \in \Pi_{q+1}$ be such that $\sigma_q = \bar{D} f_{q+1}^{-1} = \bar{D} f'^{-1}_{q+1} = \sigma'_q$. It follows that $\bar{D}(f_{q+1}^{-1} \circ f'_{q+1}) = 0$ and one can find $g \in aut(X)$ such that $f'_{q+1} = f_{q+1} \circ j_{q+1}(g)$ providing $\bar{\omega}' = f'^{-1}_q(\omega) = (f_q \circ j_q(g))^{-1}(\omega) = j_q(g)^{-1}(f_q^{-1}(\omega)) = j_q(g)^{-1}(\bar{\omega})$.

Proposition 6.3. *Natural transformations of $\mathcal{F}$ over the source in the nonlinear Janet sequence correspond to gauge transformations of $T^* \otimes R_q$ or C_1 over the target in the nonlinear Spencer sequence. Similarly, the Lie derivative $\mathcal{D}\xi = \mathcal{L}(\xi)\omega \in F_0$ in the linear Janet sequence corresponds to the Spencer operator $D\xi_{q+1} \in T^* \otimes R_q$ or $D_1\xi_q \in C_1$ in the linear Spencer sequence.*

With a slight abuse of language $\delta f = \eta \circ f \Leftrightarrow \delta f \circ f^{-1} = \eta \Leftrightarrow f^{-1} \circ \delta f = \xi$ when $\eta = T(f)(\xi)$ and we get $j_q(f)^{-1}(\omega) = \bar{\omega} \Rightarrow j_q(f + \delta f)^{-1}(\omega) = \bar{\omega} + \delta\bar{\omega}$ that is $j_q(f^{-1} \circ (f + \delta f))^{-1}(\bar{\omega}) = \bar{\omega} + \delta\bar{\omega} \Rightarrow \delta\bar{\omega} = \mathcal{L}(\xi)\bar{\omega}$ and $j_q((f + \delta f) \circ f^{-1} \circ f)^{-1}(\omega) = j_q(f)^{-1}(j_q((f + \delta f) \circ f^{-1})^{-1}(\omega)) \Rightarrow \delta\bar{\omega} = j_q(f)^{-1}(\mathcal{L}(\eta)\omega)$.
Passing to the infinitesimal point of view, we obtain the following generalization of Remark 3.3 which is important for applications ([2], AJSE-mathematics):

Corollary 6.4. $\delta\bar{\omega} = \mathcal{L}(\xi)\bar{\omega} = j_q(f)^{-1}(\mathcal{L}(\eta)\omega)$.

Example 6.2. *In Example 3.1 with $n = 1, q = 1$, we have $\omega(f(x))f_x(x) = \bar{\omega}(x), \omega(f(x))f_{xx}(x) + \partial_y\omega(f(x))f_x^2(x) = \partial_x\bar{\omega}(x)$ and obtain therefore $\omega\sigma_{y,y} + \sigma_{,y}\partial_y\omega \equiv -\omega(1/f_x)(\partial_x f_x - f_{xx})(1/\partial_x f) + ((f_x/\partial_x f) - 1)\partial_y\omega = 0$ whenever $y = f(x)$. The case of an affine stucture needs more work.*

7. Cosserat versus Weyl: New perspectives for physics

As an application of the previous mehods, let us now consider the *conformal Killing system*:

$$\hat{R}_1 \subset J_1(T) \quad \omega_{rj}\xi_i^r + \omega_{ir}\xi_j^r + \xi^r\partial_r\omega_{ij} = A(x)\omega_{ij}$$

with symbols:

$$\hat{g}_2 \subset S_2T^* \otimes T \qquad n\xi_{ij}^k = \delta_i^k\xi_{rj}^r + \delta_j^k\xi_{ri}^r - \omega_{ij}\omega^{ks}\xi_{rs}^r \qquad \Rightarrow \qquad \hat{g}_3 = 0, \forall n \geq 3$$

obtained by eliminating the arbitrary function $A(x)$, where ω is the Euclidean metric when $n = 1$ (line), $n = 2$ (plane) or $n = 3$ (space) and the Minskowskian metric when $n = 4$ (space-time).

The brothers Cosserat were only dealing with the *Killing subsystem*:

$$R_1 \subset \hat{R}_1 \qquad \qquad \omega_{rj}\xi_i^r + \omega_{ir}\xi_j^r + \xi^r\partial_r\omega_{ij} = 0$$

that is with $\{\xi^k, \xi_i^k \mid \xi_r^r = 0, \xi_{ij}^k = 0\} = \{translations, rotations\}$ when $A(x) = 0$, while, *in a somehow complementary way*, Weyl was mainly dealing with $\{\xi_r^r, \xi_{ri}^r\} = \{dilatation, elations\}$. Accordingly, one has ([7]):

Theorem 7.1. *The Cosserat equations ([C], p 137 for $n = 3$, p 167 for $n = 4$):*

$$\partial_r\sigma^{i,r} = f^i \qquad , \qquad \partial_r\mu^{ij,r} + \sigma^{i,j} - \sigma^{j,i} = m^{ij}$$

are exactly described by the formal adjoint of the first Spencer operator $D_1 : R_1 \to T^* \otimes R_1$. Introducing $\phi^{r,ij} = -\phi^{r,ji}$ and $\psi^{rs,ij} = -\psi^{rs,ji} = -\psi^{sr,ij}$, they can be parametrized by the formal *adjoint of the second Spencer operator $D_2 : T^* \otimes R_1 \to \wedge^2 T^* \otimes R_1$:*

$$\sigma^{i,j} = \partial_r \phi^{i,jr} \quad , \quad \mu^{ij,r} = \partial_s \psi^{ij,rs} + \phi^{j,ir} - \phi^{i,jr}$$

Example 7.1. *When $n = 2$, lowering the indices by means of the constant metric ω, we just need to look for the factors of ξ_1, ξ_2 and $\xi_{1,2}$ in the integration by parts of the sum:*

$$\sigma^{1,1}(\partial_1 \xi_1 - \xi_{1,1}) + \sigma^{1,2}(\partial_2 \xi_1 - \xi_{1,2}) + \sigma^{2,1}(\partial_1 \xi_2 - \xi_{2,1}) + \sigma^{2,2}(\partial_2 \xi_2 - \xi_{2,2}) + \mu^{12,r}(\partial_r \xi_{1,2} - \xi_{1,2r})$$

Finally, setting $\phi^{1,12} = \phi^1, \phi^{2,12} = \phi^2, \psi^{12,12} = \phi^3$, we obtain the nontrivial parametrization $\sigma^{1,1} = \partial_2 \phi^1, \sigma^{1,2} = -\partial_1 \phi^1, \sigma^{2,1} = -\partial_2 \phi^2, \sigma^{2,2} = \partial_1 \phi^2, \mu^{12,1} = \partial_2 \phi^3 + \phi^1, \mu^{12,2} = -\partial_1 \phi^3 - \phi^2$ in a coherent way with the Airy parametrization obtained when $\phi^1 = \partial_2 \phi, \phi^2 = \partial_1 \phi, \phi^3 = -\phi$.

Remark 7.1. *First of all, it is clear that [C] (p 13,14 for $n = 1$, p 75,76 for $n = 2$) still deals with $m = 3$ for the "ambient space", that is with the construction of the nonlinear gauge sequence, in particular for the dynamical study of a line with arc length s and time t considered as a surface, hence with no way to pass from the source to the target, only possible, as we have seen, when $m = n = 3$ by using the nonlinear Spencer sequence. For $n = 4$, the group of rigid motions of space is extended to space-time by using only a translation on time and we can rewrite the formulas in ([C], p 167) as follows:*

$$\frac{d}{dt} = \frac{dx}{dt}\frac{\partial}{\partial x} + \frac{dy}{dt}\frac{\partial}{\partial y} + \frac{dz}{dt}\frac{\partial}{\partial z} + \frac{\partial}{\partial t} \Rightarrow \frac{\partial p_{xx}}{\partial x} + \ldots + \frac{1}{\Delta}\frac{dA}{dt} = \frac{\partial}{\partial x}\left(p_{xx} + \frac{A}{\Delta}\frac{dx}{dt}\right) + \ldots + \frac{\partial}{\partial t}\left(\frac{A}{\Delta}\right)$$

$$\frac{\partial q_{xx}}{\partial x} + \ldots + p_{yz} - p_{zy} + \frac{1}{\Delta}\frac{dP}{dt} + \frac{C}{\Delta}\frac{dy}{dt} - \frac{B}{\Delta}\frac{dz}{dt} = \frac{\partial}{\partial x}\left(q_{xx} + \frac{P}{\Delta}\frac{dx}{dt}\right) + \ldots$$
$$+ \frac{\partial}{\partial t}\left(\frac{P}{\Delta}\right) + \left(p_{yz} + \frac{C}{\Delta}\frac{dy}{dt}\right) - \left(p_{zy} + \frac{B}{\Delta}\frac{dz}{dt}\right)$$

It is essential to notice that the Cosserat equations for $n = 3$ are still introduced today in a phenomenological way ([35] is a good example), contrary to the "deductive" way used in ([C], p 1-6) and that "intuition" will never allow to provide the relativistic Cosserat equations for $n = 4$ which are presented for the first time.

Theorem 7.2. *The Weyl equations ([W], §35) are exactly described by the formal adjoint of the first Spencer operator $D_1 : \hat{R}_2 \to T^* \otimes \hat{R}_2$ when $n = 4$ and can be parametrized by the formal adjoint of the second Spencer operator $D_2 : T^* \otimes \hat{R}_2 \to T^* \otimes \hat{R}_2$. In particular, among the components of the first Spencer operator, one has $\partial_i \xi^r_{rj} - \xi^r_{ijr} = \partial_i \xi^r_{rj}$ and thus the components $\partial_i \xi^r_{rj} - \partial_j \xi^r_{ri} = F_{ij}$ of the EM field with EM potential $\xi^r_{ri} = A_i$ coming from the second order jets (elations). It follows that D_1 projects onto $d : T^* \to \wedge^2 T^*$ and thus D_2 projects onto the first set of Maxwell equations described by $d : \wedge^2 T^* \to \wedge^3 T^*$. Indeed, the Spencer sequence projects onto the Poincaré sequence with a shift by $+1$ in the degree of the exterior forms involved because both sequences are made with first order involutive operators and the comment after diagram (1) can thus be used. By duality, the second set of Maxwell equations thus appears among the Weyl equations which project onto the Cosserat equations because of the inclusion $R_1 \simeq R_2 \subset \hat{R}_2$.*

Remark 7.2. *When $n = 4$, the Poincaré group (10 parameters) is a subgroup of the conformal group (15 parameters) which is not a maximal subgroup because it is a subgroup of the Weyl group (11 parameters) obtained by adding the only dilatation with infinitesimal generators $x^i \partial_i$. However, the optical group is another subgroup with 10 parameters which is maximal and the same procedure may be applied to all these subgroups in order to study coupling phenomena. It is also important to notice that* *the first and second sets of Maxwell equations are invariant by any diffeomorphism and the conformal group is only the group of invariance of the Minkowski constitutive laws in vacuum ([20])([27], p 492).*

Remark 7.3. *Though striking it may look like, there is no conceptual difference between the Cosserat and Maxwell equations on space-time. As a byproduct, separating space from time, there is no conceptual difference between the Lamé constants (mass per unit volume) of elasticity and the magnetic (dielectric) constants of EM appearing in the respective wave speeds. For example, the speed of longitudinal free vibrations of a thin elastic bar with Young modulus E and mass per unit volume ρ is $v = \sqrt{\frac{E}{\rho}}$ while the speed of light in a medium with magnetic constant μ and dielectric constant ϵ is $v = \sqrt{\frac{1/\mu}{\epsilon}}$. In the first case, we have the 1-dimensional dynamical equations:*

$$\delta \int (\frac{1}{2}E(\frac{\partial \zeta}{\partial x})^2 - \frac{1}{2}\rho(\frac{\partial \zeta}{\partial t})^2) dx dt = 0 \Rightarrow E\frac{\partial^2 \zeta}{\partial x^2} - \rho\frac{\partial^2 \zeta}{\partial t^2} = 0$$

In the second case, studying the propagation in vacuum for simplicity, one uses to set $\vec{H} = (1/\mu_0)\vec{B}, \vec{D} = \epsilon_0\vec{E}$ with $\epsilon_0\mu_0 c^2 = 1$ in the induction equations and to substitute the space-time parametrization $dA = F$ of the field equations $dF = 0$ in the variational condition $\delta \int (\frac{1}{2}\epsilon_0\vec{E}^2 - \frac{1}{2}(1/\mu_0)\vec{B}^2) dx dt = 0$. However, the second order PD equations thus obtained become wave equations only if one assumes the Lorentz condition $div(A) = \omega^{ij}\partial_i A_j = 0$ ([20]). This is not correct because the Lagrangian of the corresponding variational problem with constraint must contain the additional term $\lambda div(A)$ where λ is a Lagrange multiplier providing the equations $\Box A = d\lambda$ as a 1-form and thus $\Box F = 0$ as a 2-form when $\Box$ is the Dalembertian ([27], p 885).

Remark 7.4. *When studying static phenomena, $\epsilon = (\epsilon_{ij})$ and $\vec{E} = (E^i)$ are now on equal footing in the Lagrangian, exactly like in the technique of finite elements. Starting with a homogeneous medium at rest with no stress and electric induction, we may consider a quadratic Lagrangian $A^{ijkl}\epsilon_{ij}\epsilon_{kl} + B^{ij}E_iE_j + C^{ijk}\epsilon_{ij}E_k$ obtained by moving the indices by means of the Euclidean metric. The two first terms describe (pure) linear elasticity and electrostatic while only the last quadratic coupling term may be used in order to describe coupling phenomena. For an isotropic medium, the 3-tensor C must vanish and such a coupling phenomenon, called piezzoelectricity, can only appear in non-isotropic media like crystals, providing the additional stress $\sigma^{ij} = C^{ijk}E_k$ and/or an additional electric induction $D^k = C^{ijk}\epsilon_{ij}$. Accordingly, if the medium is fixed, for example between the plates of a condenser, an electric field may provide stress inside while, if the medium is deformed as in the piezzo-lighters, an electric induction may appear and produce a spark. Finally, for an isotropic medium, we can only add a cubic coupling term $C^{ijkl}\epsilon_{ij}E_kE_l$ responsible for photoelasticity as it provides the additional electric induction $D^l = (C^{ijkl}\epsilon_{ij})E_k$, modifying therefore the dielectric constant by a term depending linearly on the deformation and thus modifying the index of refraction n because $\epsilon\mu_0 c^2 = n^2$ with $\epsilon_0\mu_0 c^2 = 1$ in vacuum leads to $\epsilon = n^2\epsilon_0$. We may also identify the dimensionless "speed" $v^k/c \ll 1, \forall k = 1, 2, 3$ (time derivative of position) with a first jet (Lorentz rotation) by setting $\partial_4\zeta^k - \zeta_4^k = 0$ and introduce the speed of deformation by the formula $2v_{ij} = \omega_{rj}(\partial_i\zeta_4^r - \zeta_{i4}^r) + \omega_{ir}(\partial_j\zeta_4^r - \zeta_{j4}^r) = \omega_{rj}\partial_i\zeta_4^r + \omega_{ir}\partial_j\zeta_4^r = \partial_4(\omega_{rj}\partial_i\zeta^r + \omega_{ir}\partial_j\zeta^r) = \omega_{rj}\partial_i v^r + \omega_{ir}\partial_j v^r = \partial_4\epsilon_{ij}, \forall 1 \leq i, j \leq 3$ in order to obtain streaming birefringence in a similar way. These results perfectly agree with most of the field-matter couplings known in engineering sciences ([28]) but contradict gauge theory ([15],[26]) and general relativity ([W],[21]).*

In order to justify the last remark, let G be a Lie group with identity e and parameters a acting on X through the group action $X \times G \to X : (x, a) \to y = f(x, a)$ and (local) infinitesimal generators θ_τ satisfying $[\theta_\rho, \theta_\sigma] = c_{\rho\sigma}^\tau\theta_\tau$ for $\rho, \sigma, \tau = 1, ..., dim(G)$. We may prolong the *graph* of this action by differentiating q times the action law in order to eliminate the parameters in the following commutative and exact diagram where $\mathcal{R}_q$ is a Lie groupoid with local coordinates (x, y_q), *source* projection $\alpha_q : (x, y_q) \rightarrow (x)$ and *target* projection $\beta_q : (x, y_q) \rightarrow (y)$ when q is large enough:

$$
\begin{array}{ccc}
0 \rightarrow X \times G \longrightarrow & \mathcal{R}_q & \rightarrow 0 \\
\| & \alpha_q \swarrow \quad \searrow \beta_q & \\
X \times G \rightarrow & X \quad \times \quad X &
\end{array}
$$

The link between the various sections of the trivial principal bundle on the left (*gauging procedure*) and the various corresponding sections of the Lie groupoid on the right with respect to the source projection is expressed by the next commutative and exact diagram:

$$
\begin{array}{ccccc}
0 \rightarrow & X & \times G = & \mathcal{R}_q & \rightarrow 0 \\
a = cst \uparrow\downarrow\uparrow & a(x) & & j_q(f) \uparrow\downarrow\uparrow f_q & \\
& X & = & X &
\end{array}
$$

Theorem 7.3. *In the above situation, the nonlinear Spencer sequence is isomorphic to the nonlinear gauge sequence and we have the following commutative and locally exact diagram:*

$$
\begin{array}{ccccc}
X \times G \rightarrow & T^* \otimes \mathcal{G} & \overset{MC}{\rightarrow} & \wedge^2 T^* \otimes \mathcal{G} \\
\downarrow & \downarrow & & \downarrow \\
0 \rightarrow \Gamma \rightarrow & \mathcal{R}_q & \overset{\bar{D}}{\rightarrow} T^* \otimes \mathcal{R}_q & \overset{\bar{D}'}{\rightarrow} \wedge^2 T^* \otimes \mathcal{R}_q
\end{array}
$$

The action is essential in the Spencer sequence but disappears in the gauge sequence.

Proof. If we consider the action $y = f(x, a)$ and start with a section $(x) \rightarrow (x, a(x))$ of $X \times G$, we obtain the section $(x) \rightarrow (x, f^k_\mu(x) = \partial_\mu f^k(x, a(x)))$ of $\mathcal{R}_q$. Setting $b = a^{-1} = b(a)$, we get $y = f(x, a) \Rightarrow x = f(y, b) \Rightarrow y = f(f(y, b(a), a)$ and thus $\frac{\partial y}{\partial x} \frac{\partial f}{\partial b} \frac{\partial b}{\partial a} + \frac{\partial y}{\partial a} = 0$ with $\frac{\partial f}{\partial b} = \theta(x)\omega(b)$ from the first fundamental theorem of Lie. With $-\omega(b)db = -dbb^{-1} = a^{-1}da$, we obtain:

$$
\begin{aligned}
\partial_i f^k_\mu - f^k_{\mu+1_i} &= d_i(\partial_\mu f^k(x, a(x)) - \partial_{\mu+1_i} f^k(x, a(x)) \\
&= \partial_\mu(\tfrac{\partial f^k}{\partial a^\tau})\partial_i a^\tau \\
&= -\partial_\mu(\tfrac{\partial f^k}{\partial x^r}\theta^r_\tau(x))\omega^\tau_\sigma(b)\tfrac{\partial b^\sigma}{\partial a^\tau}\partial_i a^\tau
\end{aligned}
$$

and thus $\chi^k_{\mu,i}(x) = A^\tau_i(x)\partial_\mu\theta^k_\tau(x)$ from the inductive formula allowing to define $\chi_q = \bar{D}f_{q+1}$. As for the commutatitvity of the right square, we have:

$$
\partial_i\chi^k_{\mu,j} - \partial_j\chi^k_{\mu,i} - \chi^k_{\mu+1_i,j} + \chi^k_{\mu+1_j,i} = (\partial_i A^\tau_j - \partial_j A^\tau_i)\partial_\mu\theta^k_\tau
$$
$$
(\{\chi_{q+1}(\partial_i), \chi_{q+1}(\partial_j)\})^k_\mu = A^\rho_i A^\sigma_j \partial_\mu([\theta_\rho, \theta_\sigma])^k = c^\tau_{\rho\sigma} A^\rho_i A^\sigma_j \partial_\mu\theta^k_\tau.
$$

$\square$

Introducing now the Lie algebra $\mathcal{G} = T_e(G)$ and the Lie algebroid $R_q \subset J_q(T)$, namely the linearization of $\mathcal{R}_q$ at the q-jet of the identity $y = x$, we get the commutative and exact diagram:

$$
\begin{array}{ccccc}
0 \rightarrow & X & \times \mathcal{G} = & R_q & \rightarrow 0 \\
\lambda = cst \uparrow\downarrow\uparrow & \lambda(x) & & j_q(\xi) \uparrow\downarrow\uparrow \xi_q & \\
& X & = & X &
\end{array}
$$

where the upper isomorphism is described by $\lambda^\tau(x) \to \xi_\mu^k(x) = \lambda^\tau(x)\partial_\mu\theta_\tau^k(x)$ for q large enough. The unusual Lie algebroid structure on $X \times \mathcal{G}$ is described by the formula: $([\lambda, \lambda'])^\tau = c_{\rho\sigma}^\tau \lambda^\rho \lambda'^\sigma + (\lambda^\rho\theta_\rho).\lambda'^\tau - (\lambda'^\sigma\theta_\sigma).\lambda^\tau$ which is induced by the ordinary bracket $[\xi, \xi']$ on T and thus depends on the action. Applying the Spencer operator, we finally obtain $\partial_i\xi_\mu^k(x) - \xi_{\mu+1_i}^k(x) = \partial_i\lambda^\tau(x)\partial_\mu\theta_\tau^k(x)$ and the linear Spencer sequence is isomorphic to the linear gauge sequence already introduced which is no longer depending on the action as it is only the tensor product of the Poincaré sequence by $\mathcal{G}$.

Example 7.2. *Let us consider the group of affine transformations of the real line $y = a^1x + a^2$ with $n = 1, \dim(G) = 2, q = 2$, $\mathcal{R}_2$ defined by the system $y_{xx} = 0$, R_2 defined by $\xi_{xx} = 0$ and the two infinitesimal generators $\theta_1 = x\frac{\partial}{\partial x}, \theta_2 = \frac{\partial}{\partial x}$. We get $f(x) = a^1(x)x + a^2(x), f_x(x) = a^1(x), f_{xx}(x) = 0$ and thus $\chi_{,x}(x) = (1/f_x(x))\partial_x f(x) - 1 = (1/a^1(x))(x\partial_x a^1(x) + \partial_x a^2(x)) = xA_x^1(x) + A_x^2(x), \chi_{x,x}(x) = (1/f_x(x))(\partial_x f_x(x) - (1/f_x(x))\partial_x f(x)f_{xx}(x)) = (1/a^1(x))\partial_x a^1(x) = A_x^1(x), \chi_{xx,x}(x) = 0$. Similarly, we get $\xi(x) = \lambda^1(x)x + \lambda^2(x), \xi_x(x) = \lambda^1(x), \xi_{xx}(x) = 0$. Finally, integrating by part the sum $\sigma(\partial_x\xi - \xi_x) + \mu(\partial_x\xi_x - \xi_{xx})$ we obtain the dual of the Spencer operator as $\partial_x\sigma = f, \partial_x\mu + \sigma = m$ that is to say the Cosserat equations for the affine group of the real line.*

It finally remains to study GR within this framework, as it is only "added" by Weyl in an independent way and, for simplicity, we shall restrict to the linearized aspect. First of all, it becomes clear from diagram (1) that the mathematical foundation of GR is based on a confusion between the operator $\mathcal{D}_1$ (*classical curvature alone*) in the Janet sequence when $\mathcal{D}$ is the Killing operator brought to involution and the operator D_2 (*gauge curvature=curvature+torsion*) in the corresponding Spencer sequence. It must also be noticed that, according to the same diagram, the bigger is the underlying group, the bigger are the Spencer bundles while, on the contrary, the smaller are the Janet bundles depending on the invariants of the group action (deformation tensor in classical elasticity is a good example). Precisely, as already noticed in Theorem 7.2, if $G \subset \hat{G}$, the Spencer sequence for G is *contained into* the Spencer sequence for $\hat{G}$ while the Janet sequence for G projects *onto* the Janet sequence for $\hat{G}$, the best picture for understanding such a phenomenon is that of two children sitting on the ends of a beam and playing at see-saw.

Such a confusion is also combined with another one well described in ([40], p 631) by the chinese saying "*To put Chang's cap on Li's head*", namely to relate the Ricci tensor (usually obtained from the Riemann tensor by contraction of indices) to the energy-momentum tensor (space-time stress), without taking into account the previous confusion relating the gauge curvature to *rotations* only while the (classical and Cosserat) stress has only to do with *translations*. In addition, it must be noticed that *the Cosserat and Maxwell equations can be parametrized while the Einstein equations cannot be parametrized* ([29]).

In order to escape from this dilemma, let us denote by $B^2(g_q), Z^2(g_q)$ and $H^2(g_q) = Z^2(g_q)/B^2(g_q)$ the coboundary (image of the left δ), cocycle (kernel of the right δ) and cohomology bundles of the δ-sequence $T^* \otimes g_{q+1} \xrightarrow{\delta} \wedge^2T^* \otimes g_q \xrightarrow{\delta} \wedge^3T^* \otimes S_{q-1}T^* \otimes T$. It can be proved that the order of the generating CC of a formally integrable operator of order q is equal to $s + 1$ when s is the smallest integer such that $H^2(g_{q+r}) = 0, \forall r \geq s$ ([26]). As an example with $n = 3$, we let the reader prove that the second order systems $y_{33} = 0, y_{23} - y_{11} = 0, y_{22} = 0$ and $y_{33} - y_{11} = 0, y_{23} = 0, y_{22} - y_{11} = 0$ have both three second order generating CC ([30]). For the Killing system $R_1 \subset J_1(T)$ with symbol g_1, we have $F_0 = J_1(T)/R_1 = T^* \otimes T/g_1$ and the short exact sequence $0 \to g_1 \to T^* \otimes T \to F_0 \to 0$.

As $q = 1$ and $g_2 = 0 \Rightarrow g_3 = 0$ we have $s = 1$ and no CC of order 1. The generating CC of order 2 only depend on $F_1 = \omega^{-1}(\mathcal{F}_1)$ according to section 2 where F_1 is now defined by the following commutative diagram with exact columns but the first on the left and exact rows:

$$
\begin{array}{ccccccccc}
& 0 & & 0 & & 0 & & & \\
& \downarrow & & \downarrow & & \downarrow & & & \\
0 \to & g_3 & \to & S_3 T^* \otimes T & \to & S_2 T^* \otimes F_0 & \to F_1 \to 0 \\
& \downarrow \delta & & \downarrow \delta & & \downarrow \delta & & \\
0 \to & T^* \otimes g_2 & \to & T^* \otimes S_2 T^* \otimes T & \to & T^* \otimes T^* \otimes F_0 & \to 0 \\
& \downarrow \delta & & \downarrow \delta & & \downarrow \delta & & \\
0 \to & \wedge^2 T^* \otimes g_1 & \to & \wedge^2 T^* \otimes T^* \otimes T & \to & \wedge^2 T^* \otimes F_0 & \to 0 \\
& \downarrow \delta & & \downarrow \delta & & \downarrow & & \\
0 \to & \wedge^3 T^* \otimes T & = & \wedge^3 T^* \otimes T & \to & 0 \\
& \downarrow & & \downarrow & & & & \\
& 0 & & 0 & & & &
\end{array}
$$

It follows from a chase([26], p 55)([27], p 192)([32], p 171) that there is a short exact *connecting sequence* $0 \to B^2(g_1) \to Z^2(g_1) \to F_1 \to 0$ leading to an isomorphism $F_1 \simeq H^2(g_1)$. The Riemann tensor is thus a section of $Riemann = F_1 = H^2(g_1) = Z^2(g_1)$ in the Killing case with $dim(Riemann) = (n^2(n+1)^2/4) - (n^2(n+1)(n+2)/6) = (n^2(n-1)^2/4) - (n^2(n-1)(n-2)/6) = n^2(n^2-1)/12$ by using either the upper row or the left column and *we find back the two algebraic properties of the Riemann tensor without using indices.*

However, for the conformal Killing system, we still have $q = 1$ but the situation is much more delicate because $g_3 = 0$ for $n \geq 3$ and $H^2(\hat{g}_2) = 0$ only for $n \geq 4$ ([26], p 435). Hence, setting similarly $\hat{F}_0 = T^* \otimes T/\hat{g}_1$, the Weyl tensor is a section of $Weyl = \hat{F}_1 = H^2(\hat{g}_1) \neq Z^2(\hat{g}_1)$. The inclusion $g_1 \subset \hat{g}_1$ and the relations $g_2 = 0, \hat{g}_3 = 0$ finally induce the following *crucial* commutative and exact *diagram* (2) ([25], p 430):

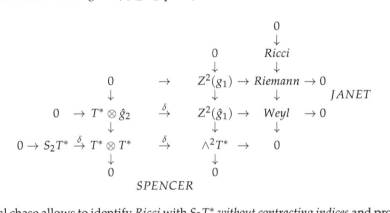

SPENCER

A diagonal chase allows to identify *Ricci* with $S_2 T^*$ *without contracting indices* and provides the splitting of $T^* \otimes T^*$ into $S_2 T^*$ *(gravitation)* and $\wedge^2 T^*$ *(electromagnetism)* in the lower horizontal sequence *obtained by using the Spencer sequence*, solving thus an old conjecture. However, $T^* \otimes T^* \simeq T^* \otimes \hat{g}_2$ has only to do with second order jets *(elations)* and not a word is left from the standard approach to GR. In addition, we obtain the following important theorem explaining for the first time classical results in an intrinsic way:

Theorem 7.4. *There exist canonical splittings of the various δ-maps appearing in the above diagram which allow to split the vertical short exact sequence on the right.*

Proof. We recall first that a short exact sequence $0 \to M' \xrightarrow{f} M \xrightarrow{g} M'' \to 0$ of modules *splits*, that is $M \simeq M' \oplus M''$, if and only if there exists a map $u : M \to M'$ with $u \circ f = id_{M'}$ or a map $v : M'' \to M$ with $g \circ v = id_{M''}$ ([3], p 73)([32], p 33). Hence, starting with $(\tau^k_{li,j}) \in T^* \otimes \hat{g}_2$, we may introduce $(\rho^k_{l,ij} = \tau^k_{li,j} - \tau^k_{lj,i}) \in B^2(\hat{g}_1) \subset Z^2(\hat{g}_1) \subset \wedge^2 T^* \otimes \hat{g}_1$ but now $\varphi_{ij} = \rho^r_{r,ij} = \tau^r_{ri,j} - \tau^r_{rj,i} = \rho_{ij} - \rho_{ji} \neq 0$ with $\rho_{ij} = \rho^r_{i,rj}$ because we have $\rho^k_{l,ij} + \rho^k_{i,jl} + \rho^k_{j,li} = 0$. With $\tau = \omega^{ij}\tau^r_{ri,j}$ and $\rho = \omega^{ij}\rho_{ij}$, we obtain $(n-2)\tau^r_{ri,j} = (n-1)\rho_{ij} + \rho_{ji} - (n/2(n-1))\omega_{ij}\rho$ and thus $n\rho = 2(n-1)\tau$. The lower sequence splits with $\varphi_{ij} \to \tau_{ij} = \tau^r_{ri,j} = (1/2)\varphi_{ij} \to \tau_{ij} - \tau_{ji} = \varphi_{ij}$ and $\rho_{ij} = \rho_{ji} \Leftrightarrow \varphi_{ij} = 0$ in $Z^2(g_1) \subset \wedge^2 T^* \otimes g_1$. It follows from a chase that the kernel of the canonical projection *Riemann* $\to$ *Weyl* is defined by $\rho^k_{l,ij} = \tau^k_{li,j} - \tau^k_{lj,i}$ with $(\rho^k_{l,ij}) \in Z^2(g_1) \subset Z^2(\hat{g}_1)$ and $(\tau^k_{li,j}) \in T^* \otimes \hat{g}_2$. Accordingly $(n-2)\tau_{ij} = n\rho_{ij} - (n/2(n-1))\omega_{ij}\rho$ provides the isomorphism *Ricci* $\simeq S_2 T^*$ and we get $n\rho^k_{l,ij} = \delta^k_i \tau_{lj} - \delta^k_j \tau_{li} + \omega_{lj}\omega^{ks}\tau_{si} - \omega_{li}\omega^{ks}\tau_{sj}$, that is:

$$\rho^k_{l,ij} = \frac{1}{(n-2)}(\delta^k_i \rho_{lj} - \delta^k_j \rho_{li} + \omega_{lj}\omega^{ks}\rho_{si} - \omega_{li}\omega^{ks}\rho_{sj}) - \frac{1}{(n-1)(n-2)}(\delta^k_i \omega_{lj} - \delta^k_j \omega_{li})\rho$$

We check that $\rho^r_{i,rj} = \rho_{ij}$, obtaining therefore a splitting of the right vertical sequence in the last diagram that allows to define the Weyl tensor by difference. These purely algebraic results only depend on ω independently of any conformal factor.   □

Example 7.3. *The free movement of a body in a constant static gravitational field $\vec{g}$ is described by $\frac{d\vec{x}}{dt} - \vec{v} = 0$, $\frac{d\vec{v}}{dt} - \vec{g} = 0$, $\frac{\partial\vec{g}}{\partial x^i} - 0 = 0$ where the "speed" is considered as a first order jet (Lorentz rotation) and the "gravity" as a second order jet (elation). Hence an accelerometer merely helps measuring the part of the Spencer operator dealing with second order jets (equivalence principle). As a byproduct, the difference $\partial_4 f^k_4 - f^k_{44}$ under the constraint $\partial_4 f^k - f^k_4$ identifying the "speed" with a first order jet allows to provide a modern version of the Gauss principle of least constraint where the extremum is now obtained with respect to the second order jets and not with respect to the "acceleration" as usual ([1], p 470). The corresponding infinitesimal variational principle $\delta \int (\rho(\partial_4 \xi^4 - \xi^4_4) + g^i(\partial_i \xi^r_r - \xi^r_{ri}) + g^{ij}(\partial_i \xi^r_{rj} - 0))dx = 0$ provides the Poisson law of gravitation with $\rho = cst$ and $\vec{g} = (g^i)$ when $g^{ij} = \lambda\omega^{ij} \Rightarrow g_i = -\partial_i\lambda$. The last term of this gravitational action in vacuum is thus of the form $\lambda div(A)$, that is exactly the term responsible for the Lorentz constraint in Remark 7.6.*

8. Conclusion

In continuum mechanics, the classical approach is based on differential invariants and only involves derivatives of finite transformations. Accordingly, the corresponding variational calculus can only describe forces as it only involves translations. It has been the idea of E. and F. Cosserat to change drastically this point of view by considering a new differential geometric tool, now called Spencer sequence, and a corresponding variational calculus involving *both* translations and rotations in order to describe torsors, that is *both* forces and couples.

About at the same time, H. Weyl tried to describe electromagnetism and gravitation by using, *in a similar but complementary way*, the dilatation and elations of the conformal group of space-time. We have shown that the underlying Spencer sequence has additional terms, *not known today*, wich explain in a unique way all the above results and the resulting field-matter couplings.

In gauge theory, the structure of electromagnetism is coming from the unitary group $U(1)$, the unit circle in the complex plane, which is *not* acting on space-time, as the *only* possibility to obtain a pure 2-form from $\wedge^2 T^* \otimes \mathcal{G}$ is to have $dim(\mathcal{G}) = 1$. However, we have explained the structure of electromagnetism from that of the conformal group of space-time, with a *shift by one step* in the interpretation of the Spencer sequence involved because the "*fields*" are now sections of $C_1 \simeq T^* \otimes \mathcal{G}$ parametrized by D_1 and thus killed by D_2.

In general relativity, we have similarly proved that the standard way of introducing the Ricci tensor was based on a *double confusion* between the Janet and Spencer sequences described by *diagrams* (1) *and* (2). In particular we have explained why the intrinsic structure of this tensor *necessarily* depends on the difference existing between the Weyl group and the conformal group which is coming from second order jets, relating for the first time on equal footing electromagnetism and gravitation to the Spencer δ-cohomology of various symbols.

Accordingly, paraphrasing W. Shakespeare, we may say:

" TO ACT OR NOT TO ACT, THAT IS THE QUESTION "

and hope future will fast give an answer !.

9. References

[1] P. Appell: Traité de Mécanique Rationnelle, Gauthier-Villars, Paris, 1909. Particularly t II concerned with analytical mechanics and t III with a Note by E. and F. Cosserat "Note sur la théorie de l'action Euclidienne", 557-629.
[2] V. Arnold: Méthodes mathématiques de la mécanique classique, Appendice 2 (Géodésiques des métriques invariantes à gauche sur des groupes de Lie et hydrodynamique des fluides parfaits), MIR, moscow, 1974,1976. (For more details, see also: J.-F. POMMARET: Arnold's hydrodynamics revisited, AJSE-mathematics, 1, 1, 2009, 157-174).

[3] I. Assem: Algèbres et Modules, Masson, Paris, 1997.
[4] G. Birkhoff: Hydrodynamics, Princeton University Press, Princeton, 1954. French translation: Hydrodynamique, Dunod, Paris, 1955.
[5] E. Cartan: Sur une généralisation de la notion de courbure de Riemann et les espaces à torsion, C. R. Académie des Sciences Paris, 174, 1922, 437-439, 593-595, 734-737, 857-860.
[6] E. Cartan: Sur les variétés à connexion affine et la théorie de la relativité généralisée, Ann. Ec. Norm. Sup., 40, 1923, 325-412; 41, 1924, 1-25; 42, 1925, 17-88.
[7] O. Chwolson: Traité de Physique (In particular III, 2, 537 + III, 3, 994 + V, 209), Hermann, Paris, 1914.
[8] E. Cosserat, F. Cosserat: Théorie des Corps Déformables, Hermann, Paris, 1909.
[9] J. Drach: Thèse de Doctorat: Essai sur une théorie générale de l'intégration et sur la classification des transcendantes, in Ann. Ec. Norm. Sup., 15, 1898, 243-384.
[10] L.P. Eisenhart: Riemannian Geometry, Princeton University Press, Princeton, 1926.
[11] H. Goldschmidt: Sur la structure des équations de Lie, J. Differential Geometry, 6, 1972, 357-373 and 7, 1972, 67-95.

[12] H. Goldschmidt, D.C. Spencer: On the nonlinear cohomology of Lie equations, I+II, Acta. Math., 136, 1973, 103-239.

[13] M. Janet: Sur les systèmes aux dérivées partielles, Journal de Math., 8, (3), 1920, 65-151.

[14] E.R. Kalman, Y.C. YO, K.S. Narenda: Controllability of linear dynamical systems, Contrib. Diff. Equations, 1 (2), 1963, 189-213.

[15] S. Kobayashi, K. Nomizu: Foundations of Differential Geometry, Vol I, J. Wiley, New York, 1963, 1969.

[16] G. Koenig: Leçons de Cinématique (The Note "Sur la cinématique d'un milieu continu" by E. Cosserat and F. Cosserat has rarely been quoted), Hermann, Paris, 1897, 391-417.

[17] E.R. Kolchin: Differential Algebra and Algebraic Groups, Academic Press, New York, 1973.

[18] A. Kumpera, D.C. Spencer: Lie Equations, Ann. Math. Studies 73, Princeton University Press, Princeton, 1972.

[19] E. Kunz: Introduction to Commutative Algebra and Algebraic Geometry, Birkhäuser, 1985.

[20] V. Ougarov: Théorie de la Relativité Restreinte, MIR, Moscow, 1969; french translation, 1979.

[21] W. Pauli: Theory of Relativity, Pergamon Press, London, 1958.

[22] H. Poincare: Sur une forme nouvelle des équations de la mécanique, C. R. Académie des Sciences Paris, 132 (7), 1901, 369-371.

[23] J.-F. Pommaret: Systems of Partial Differential Equations and Lie Pseudogroups, Gordon and Breach, New York, 1978; Russian translation: MIR, Moscow, 1983.

[24] J.-F. Pommaret: Differential Galois Theory, Gordon and Breach, New York, 1983.

[25] J.-F. Pommaret: Lie Pseudogroups and Mechanics, Gordon and Breach, New York, 1988.

[26] J.-F. Pommaret: Partial Differential Equations and Group Theory, Kluwer, Dordrecht, 1994.

[27] J.-F. Pommaret: Partial Differential Control Theory, Kluwer, Dordrecht, 2001.

[28] J.-F. Pommaret: Group interpretation of Coupling Phenomena, Acta Mechanica, 149, 2001, 23-39.

[29] J.-F. Pommaret: Parametrization of Cosserat equations, Acta Mechanica, 215, 2010, 43-55.

[30] J.-F. Pommaret: Macaulay inverse systems revisited, Journal of Symbolic Computation, 46, 2011, 1049-1069.

[31] J.F. Ritt: Differential Algebra, Dover, 1966.

[32] J. J. Rotman: An Introduction to Homological Algebra, Academic Press, 1979.

[33] D. C. Spencer: Overdetermined Systems of Partial Differential Equations, Bull. Am. Math. Soc., 75, 1965, 1-114.

[34] I. Stewart: Galois Theory, Chapman and Hall, 1973.

[35] P.P. Teodorescu: Dynamics of Linear Elastic Bodies, Editura Academiei, Bucuresti, Romania; Abacus Press, Tunbridge, Wells, 1975.

[36] E. Vessiot: Sur la théorie des groupes infinis, Ann. Ec. Norm. Sup., 20, 1903, 411-451.

[37] E. Vessiot: Sur la théorie de Galois et ses diverses généralisations, Ann. Ec. Norm. Sup., 21, 1904, 9-85.

[38] C. Yang: Magnetic monopoles, fiber bundles and gauge fields, Ann. New York Acad. Sciences, 294, 1977, 86.

[39] C.N. Yang, R.L. Mills: Conservation of isotopic gauge invariance, Phys. Rev., 96, 1954, 191-195.

[40] Z. Zou, P. Huang, Y. Zhang, G. Li: Some researches on gauge theories of gravitation, Scientia Sinica, XXII, 6, 1979, 628-636.

2

Continuum Mechanics of Solid Oxide Fuel Cells Using Three-Dimensional Reconstructed Microstructures

Sushrut Vaidya and Jeong-Ho Kim[*]
Department of Civil and Environmental Engineering, University of Connecticut,
USA

1. Introduction

A solid oxide fuel cell (SOFC) is a device that converts the chemical energy of fuels into electrical energy (Singhal & Kendall, 2003). SOFCs have received much attention from researchers due to their promise of delivering relatively clean energy at high efficiencies (Singhal & Kendall, 2003). An SOFC consists of a few basic parts: an anode, a cathode, an electrolyte, and interconnect wires (Singhal & Kendall, 2003). The electrolyte in an SOFC is a solid oxide such as Yttria-Stabilized Zirconia (YSZ). The porous anode is usually a ceramic-metal composite (so called cermet) such as the nickel-zirconia cermet (Ni-YSZ). The porous cathode is usually a composite of strontium-doped lanthanum manganite (LSM) and Yttria-Stabilized Zirconia (LSM-YSZ) (Singhal & Kendall, 2003) or a composite such as gadolinium-doped ceria-lanthanum strontium cobalt ferrite (GDC-LSCF) (Anandakumar et al., 2010). Oxygen atoms undergo reduction on the porous cathode surface, and the resulting oxide ions are transported through the electrolyte to the porous anode. Here, the oxide ions react with the fuel (such as hydrogen). Hydrogen is oxidized, and the electrons of the oxide ions are liberated. The free electrons give rise to electric current (Singhal & Kendall, 2003).

Research on SOFCs has concentrated on many different aspects, including anode, cathode, and electrolyte materials; investigating the behavior of different SOFC configurations; modeling and simulating electrochemical, thermal, and flow phenomena; and performing thermal stress and probability of failure analyses. Researchers have employed experimental, analytical, and computational approaches in their investigations. For example, Selcuk and Atkinson (1997, 2000) conducted a number of experimental studies to estimate various mechanical properties of SOFC ceramic materials such as YSZ and NiO-YSZ. They determined the biaxial flexural strength and fracture toughness of YSZ both at room temperature and an operating temperature of 900⁰C (Selcuk & Atkinson, 2000). They also experimentally studied the dependence of the Young's modulus, shear modulus, and Poisson's ratio of YSZ and NiO-YSZ (amongst other ceramic materials) on porosity (Selcuk & Atkinson, 1997). The results of these studies were summarized by the authors (Atkinson & Selcuk, 2000) where they also suggested techniques for improving the mechanical

---

[*] Corresponding Author

behavior of SOFC ceramic materials under certain operating conditions. Toftegaard et al. (2009) conducted uniaxial tensile tests on pure YSZ specimens and YSZ specimens coated with porous NiO-YSZ layers. They heat-treated the coated YSZ specimens at various temperatures to study the effect of heat treatment at different temperatures on the strength. Pihlatie et al. (2009) experimentally determined the Young's modulus (amongst other mechanical properties) of Ni-YSZ and NiO-YSZ composites as a function of porosity using the Impulse Excitation Technique (IET). They also used IET to study the dependency of the Young's modulus of these materials on temperature. Giraud and Canel (2008) also conducted experimental studies using IET to determine the variation of the Young's modulus of YSZ, LSM, and Ni-YSZ with temperature. Wilson and Barnett (2008) conducted experimental studies on Ni-YSZ/YSZ/LSM-YSZ SOFCs with Ni-YSZ anodes of different compositions to investigate the effect of varying composition of the anodes on their performance and microstructure. Their studies involved serial-sectioning using a focused ion beam scanning electron microscope (FIB-SEM) to obtain images of the microstructures of the different samples. They conducted stereological analyses on these images to calculate volume fractions and triple-phase boundary (TPB) densities for their samples. Zhang et al. (2008) proposed an analytical model for calculating residual stresses in a single SOFC with NiO-YSZ/YSZ/LSM composition. They used their model to estimate the residual stresses in an SOFC at room temperature and to study the variation of the stresses in the different components with changes in component thicknesses. They also carried out a Weibull analysis to calculate the probability of failure of the anode, and they studied the variation of the failure probability of the anode with changes in component thicknesses.

Laurencin et al. (2008) have proposed a numerical (finite element analysis-based) tool for studying the degradation of anode-supported and electrolyte-supported planar SOFCs under several types of mechanical loads, including residual stresses. They have also calculated the failure probabilities of the SOFCs using Weibull analysis. Pitakthapanaphong and Busso (2005) carried out finite element analyses to investigate the fracture of multi-layered systems used in SOFCs, such as LSM films on a YSZ substrate. They have pointed out that fracture is caused by large residual stresses induced during the SOFC manufacturing process due to thermal expansion coefficient (TEC) mismatch between different layers. They observed different cracking patterns (surface cracks, channeling cracks, and interfacial cracks) in physical samples of multi-layered systems. Their study involved FE simulations to determine the crack driving force (energy release rate) for the three observed cracking patterns. Johnson and Qu (2008) used a three-dimensional stochastic reconstruction method to create multiple realizations of the microstructure of porous Ni-YSZ cermet used as SOFC anode material. They analyzed these microstructure realizations using finite element software to determine the effective elastic modulus and effective coefficient of thermal expansion (CTE) of Ni-YSZ as a function of temperature. Anandakumar et al. (2010) carried out FE analyses to estimate thermal stresses and probability of failure in functionally graded SOFCs. They employed a continuum mechanics approach and used graded finite elements to discretize effective media consisting of NiO-YSZ/YSZ/LSM as well as NiO-YSZ/YSZ/GDC-LSCF. They used the Weibull method to determine the failure probability of the individual components of the SOFC, as well as the failure probability of the whole SOFC. They found that the thermal stresses developed in functionally graded SOFCs under spatially uniform and non-uniform temperature loads are lower than those induced in conventional layered SOFCs. They also found that functionally graded SOFCs show a lower probability of failure than other types of SOFCs.

In this work, three-dimensional micromechanical finite element (FE) models for real solid oxide fuel cell (SOFC) anode and cathode microstructures are generated from a stack of two-dimensional image-based FE models of anode and cathode microstructures. Finite element analysis (FEA) of the models is carried out to determine their mechanical response to a steady-state temperature change from room temperature up to an operating temperature. The resulting stress distribution is determined in each case, and the stresses are analyzed using the Weibull method to calculate the probability of failure. The anode material is Ni-YSZ, while the cathode material is LSM-YSZ. Both linear elastic and elastic-plastic (nonlinear) behaviors are considered for nickel in the analysis of the anode. It is observed that the linear elastic models underestimate the probability of failure of the anode. The effect of temperature-dependent material properties on the probability of failure of the anode and cathode is also investigated. *The novelties of this work include micromechanical finite element analysis of the mechanical response of anode and cathode microstructural models considering temperature-dependent material properties and nonlinear elastic-plastic behavior of the nickel phase.*

2. Image-based microstructural finite element models

The first step in this work was to digitally reconstruct a three-dimensional (3-D) anode microstructure from two-dimensional (2-D) images of anode cross-sections obtained using focused ion beam-scanning electron microscopy (FIB-SEM). The 2-D images of the anode and cathode microstructures were obtained from Dr. Scott Barnett's research group at Northwestern University (Wilson et al., 2006, 2009). The initial 3-D reconstruction was achieved using IMOD (Kramer et al., 1996), a free collection of image processing programs developed by scientists at the Boulder Laboratory for 3-D Electron Microscopy of Cells. IMOD is capable of creating a stack of 2-D images, interpolating the gaps between consecutive images, and creating and displaying the 3-D model. A few representative 2-D images and the 3-D reconstruction of the anode microstructure are shown in Figure 1. The in-plane dimensions of the reconstructed anode are 5 μm x 6 μm, while the thickness is 3.54 μm.

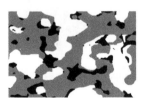 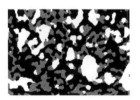 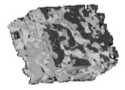

Fig. 1. Two-dimensional SEM images of anode (left) and cathode (center) cross-sections (Wilson et al., 2006, 2009) and a reconstructed three-dimensional anode (right)

In the SEM images of the anode, white (pixel value 255) corresponds to nickel, gray (pixel value 127) to YSZ, and black (pixel value 0) to the pores. In the images of the cathode, white (pixel value 255) represents LSM, gray (pixel value 127) represents YSZ, and black (pixel value 0) represents the pores. The reconstructed 3-D model was used as a check on the geometry of the 3-D FE model.

The next step involved the creation of a single 2-D FE model from a single 2-D SEM image. FE modeling was carried out using the commercial FE software ABAQUS v6.9 (Dassault Systems Simulia Corp., Providence, Rhode Island, USA). This was done by writing a MATLAB ® program (R2010a, The MathWorks, Inc., Natick, Massachusetts, USA) to recreate the geometry of the image using 2-D finite elements (4-node quadrilateral elements) and write the geometry data to an ABAQUS input file. Exactly one element was assigned to each pixel in the image, and the element was assigned to the appropriate element set (nickel or YSZ) based on the pixel value. Information concerning the material properties, boundary conditions, initial temperature, temperature field, and required outputs (e.g. principal stresses) was also specified in the input file. The input file was then run using ABAQUS to generate the 2-D FE model as shown in Figure 2.

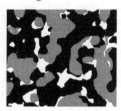

Fig. 2. Two-dimensional FE model of a single cross-section of the SOFC anode

The 3-D FE anode and cathode models were created by making a stack of all the 2-D images and introducing a "buffer" plane between each pair of consecutive images. This was necessary and useful to ensure a simple step variation in material properties between corresponding regions in two consecutive images. Then the gaps between consecutive images were interpolated by assigning one three-dimensional 8-node brick element to each volumetric pixel (or voxel). Thus, the 3-D geometries of the anode and cathode microstructures were recreated in the 3-D FE models of the anode and cathode. Various free-body cuts of the 3-D FE anode model are shown in Figure 3.

Fig. 3. Free-body cuts of the three-dimensional FE model of the SOFC anode

3. Finite element analysis of anode and cathode models

3.1 Analysis models and metrics

The FE analysis of the anode and cathode models was carried out to investigate the effect of various temperature loads, as well as the effect of variation of material properties with temperature, on the mechanical response and probability of failure. In the case of the anode, the effect of nonlinear (elastic-plastic) behavior versus linear elastic behavior of nickel was also investigated, which has not been studied in the literature.

The FE analyses of the anode and cathode models were divided into different categories as explained in Tables 1 and 2. In each case, the FE model was subjected to fixed boundary conditions (i.e. all nodes on each of the six faces were allowed neither to translate nor to rotate). The behavior of the model with increasing temperature loads was investigated by subjecting the model to eight different spatially uniform predefined temperature fields of magnitude 120°C, 220°C, 320°C, ..., 820°C. In each analysis, the initial temperature was specified as 20°C (room temperature), so that the model was subjected to eight different magnitudes of temperature change ($\Delta T = 100°C, 200°C, 300°C, ..., 800°C$)..

Case	Ni	YSZ	Temperature-dependence (Ni)	Temperature-dependence (YSZ)
Case 1: Temperature-independent	Linear elastic	Linear elastic	None	None
Case 2: Temperature-dependent CTEs	Linear elastic	Linear elastic	CTE	CTE
Case 3: Elastic-plastic behavior of Ni	Elastic-plastic	Linear elastic	CTE	Young's modulus, CTE

Table 1. Metrics for finite element analyses of anode

Case	LSM	YSZ	Temperature-dependence (LSM)	Temperature-dependence (YSZ)
Case 1: Temperature-independent	Linear elastic	Linear elastic	None	None
Case 2: Temperature-dependent	Linear elastic	Linear elastic	Young's modulus	Young's modulus, CTE

Table 2. Metrics for finite element analyses of cathode

3.2 Material properties

Table 3 lists the room temperature material properties used for nickel, YSZ and LSM (Johnson & Qu, 2008; Anandakumar et al., 2010).

Material	Young's modulus (GPa)	Poisson's ratio	CTE (10^{-6} °C^{-1})
Nickel	207	0.31	12.50
YSZ	205	0.30	10.40
LSM	40	0.25	11.40

Table 3. Room temperature material properties used in FE analyses

Figure 4 shows the variation of the coefficients of thermal expansion of nickel and YSZ with temperature (Johnson & Qu, 2008). The CTE of LSM was assumed to be constant over the temperature range considered. The room temperature value of the CTE of LSM (as shown in Table 3) was used in the FE analyses of the cathode.

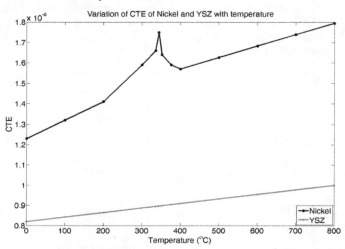

Fig. 4. Variation of CTE of nickel and YSZ with temperature (Johnson & Qu, 2008).

Figure 5 shows the variation of the Young's modulus of LSM and YSZ with temperature (Giraud & Canel, 2008). The Young's modulus of nickel was assumed to be constant over the temperature range considered. The room temperature value of the Young's modulus of nickel (as shown in Table 3) was used in the FE analyses of the anode.

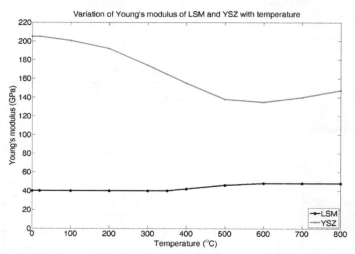

Fig. 5. Variation of Young's modulus of LSM and YSZ with temperature (Giraud and Canel, 2008).

Figure 6 shows the stress-strain curve used to describe the elastic-plastic behavior of nickel (Ebrahimi et al., 1999). It was assumed in this work that the stress-strain curve of nickel does not change over the temperature range considered.

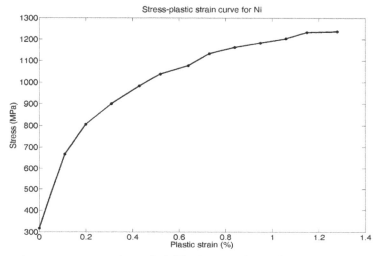

Fig. 6. Stress-plastic strain curve for nickel (Ebrahimi et al., 1999).

4. Finite element analysis results and discussion

4.1 Stress analysis

The full 3-D FE models of the anode (50:50 NiO:YSZ weight percentage composition) and cathode (50:50 LSM:YSZ weight percentage composition) are shown in Figure 7. The FE anode model consists of 406,465 elements and 473,181 nodes. The cathode model has 244,584 elements and 395,131 nodes.

Fig. 7. Three-dimensional FE models of anode (left) and cathode (right)

4.1.1 Anode microstructure

The von Mises stress contour plots for the anode at $\Delta T = 100^\circ C$, $500^\circ C$, and $800^\circ C$ are shown considering elastic-plastic behavior of nickel in Figure 8. The stress values are in units of N/m^2 (i.e., Pa).

Fig. 8. Von Mises stress contour plots for anode considering elastic-plastic behavior of nickel: (left to right) $\Delta T = 100^0C$, $\Delta T = 500^0C$, $\Delta T = 800^0C$

Figure 8 shows that as ΔT increases from 100^0C to 800^0C, the stresses in the anode also increase. This happens because thermal stress is proportional to the CTE, and the CTEs of both nickel and YSZ increase with temperature, as seen from Figure 4. Also, the stress plots show that the stresses are greater near the regions of pores due to stress concentration, as expected. Similar results are obtained for the cases with temperature-independent material properties and temperature-dependent CTEs. The effect of the elastic-plastic behavior of nickel on the principal tensile stress values (as compared with the linear elastic behavior assumed in the cases with temperature-independent material properties and temperature-dependent CTEs) is discussed in section 4.2.1, which deals with failure probability calculations for the anode.

4.1.2 Cathode microstructure

The von Mises stress contour plots for the cathode at $\Delta T = 100^0C$, 500^0C, and 800^0C are shown in Figure 9 considering temperature-dependent material properties.

Fig. 9. Von Mises stress contour plots for cathode considering temperature-dependent material properties: (left to right) $\Delta T = 100^0C$, $\Delta T = 500^0C$, $\Delta T = 800^0C$

Figure 9 shows that as ΔT increases from 100^0C to 800^0C, the stresses in the cathode also increase. This result can be explained, just as in the case of the anode, by the fact that thermal stress is proportional to the CTE, and the CTE of YSZ increases with temperature while the CTE of LSM is assumed constant over the temperature range considered. Also, the plots show that the stresses are greater near the regions of pores due to stress concentration. Similar results were obtained for the case with temperature-independent material properties. The effect of temperature-independent versus temperature-dependent material properties on the principal tensile stresses induced in the cathode is discussed in section 4.2.2, which deals with failure probability calculations for the cathode.

4.2 Probability of failure analysis

Ceramic materials exhibit brittle behavior under tensile stress. Also, unlike metals, they show wide variability in tensile strength values and follow a statistical strength distribution. Thus, the Weibull method of analysis (Weibull, 1951; Laurencin et al., 2008) was used to calculate the probability of failure of each SOFC component (anode/cathode). According to the Weibull method, the survival probability of a particular component j under the action of a tensile stress σ is given by (Laurencin et al., 2008):

$$P_s^j(\sigma, V_j) = \exp\left(-\int_{V_j}\left(\frac{\sigma}{\sigma_0}\right)^m \frac{dV_j}{V_0}\right)$$

(1)

where j = anode or cathode, V_j is the volume of component j, V_0 is a characteristic specimen volume (reference volume) for the material of component j, σ_0 is the characteristic strength of the material of component j, and m is the Weibull modulus of the material. The characteristic strength σ_0 is also the scale parameter for the distribution, while the Weibull modulus m is the shape parameter. The reference volume V_0 is related to the characteristic strength σ_0 of the material.

In our case, however, the Weibull method was slightly modified to account for the fact that the anode and cathode materials are composites made up of two different components (Ni-YSZ for the anode and LSM-YSZ for the cathode). The method employed is described next. The Weibull parameters used for the ceramic materials (LSM and YSZ) are shown in Table 4 (Laurencin et al., 2008). Only room temperature values of the Weibull parameters were used in this study.

Material	Weibull modulus, m	Characteristic strength, σ_0 (MPa)	Reference volume, V_0 (mm³)
LSM	7.0	52.0	1.21
YSZ	7.0	446.0	0.35

Table 4. Weibull parameters of ceramic materials (room temperature values)

The results of each stress analysis case were post-processed by writing programs to extract the three principal stress values from each element in the anode and cathode FE models. These principal stresses were then used to perform a Weibull analysis to determine the probability of failure of the anode and cathode at each ΔT value. Since the SOFC component materials are subjected to a multi-axial state of stress, the total survival probability of each ceramic phase of the anode/cathode under the action of the three principal stresses (σ_1, σ_2, and σ_3) was calculated. The principal stresses were assumed to act independently, and the total survival probability was calculated as the product of the survival probabilities under the action of each individual principal stress (Laurencin et al., 2008):

$$P_s^j(\overline{\sigma}, V_j) = \prod_{i=1}^{3} P_s^j(\sigma_i, V_j)$$

(2)

Also,

$$P_s^j(\sigma_i, V_j) = \exp\left(-\int_{V_j}\left(\frac{\sigma_i}{\sigma_0}\right)^m \frac{dV_j}{V_0}\right) \qquad (3)$$

where, j = YSZ for the anode, j = LSM or YSZ for the cathode, and i = 1, 2, and 3. Only tensile values of the three principal stresses were used in the Weibull analysis. The probability of failure of each phase was then calculated as follows (Anandakumar et al., 2010):

$$P_f = 1.0 - P_s^j(\overline{\sigma}, V_j) \qquad (4)$$

The probability of failure of the anode was calculated as the failure probability of the YSZ phase, keeping in mind that the anode material is a cermet composite (Ni-YSZ), and that the Weibull distribution is more appropriate for calculating the failure probability of ceramics (such as YSZ) (Laurencin et al., 2008). The strength distribution for metals such as nickel is closer to a normal distribution (Meyers & Chawla, 1999). Since the cathode is a composite of two different ceramic materials (LSM-YSZ), the probability of failure of the cathode was calculated by extracting positive (tensile) values of the three principal stresses from each element in the LSM and YSZ element sets of the cathode FE model, and subjecting these to the Weibull analysis. This resulted in two different failure probability values for the LSM and YSZ phases of the cathode, which were combined into a single probability of failure value for the cathode by assuming that the cathode fails when either phase fails (or when both phases fail simultaneously). The probability that both phases fail simultaneously was calculated by assuming that the failures of the two phases are independent events, and hence the probability of simultaneous failure of the two phases is just the product of the probabilities of failure of LSM and YSZ:

$$P_f^{cathode} = P_f(LSM \cup YSZ)$$

$$\Rightarrow P_f^{cathode} = P_f(LSM) + P_f(YSZ) - P_f(LSM \cap YSZ)$$

$$\Rightarrow P_f^{cathode} = P_f(LSM) + P_f(YSZ) - P_f(LSM)P_f(YSZ)$$

4.2.1 Anode

The probability of failure (P_f) value for the YSZ phase of the anode was calculated at each ΔT value (100°C, 200°C, ..., 800°C) for each case described in Table 1. These values are plotted in Figure 10. Since these P_f values are calculated on the basis of the tensile principal stresses in the YSZ phase, the variation of the maximum principal tensile stress (MPTS) in the YSZ phase of the anode with temperature in all three cases (temperature-independent material properties, temperature-dependent CTEs, and elastic-plastic behavior of Ni) is shown in Figure 11.

The P_f plot for the anode (Figure 10) shows that the probability of failure increases with increasing ΔT values (and hence increasing stresses), for each of the three cases considered (temperature-independent material properties, temperature-dependent CTEs, and elastic-plastic behavior of nickel). Also, the plot shows that the linear elastic material behavior models significantly underestimate the probability of failure of the anode (defined by the P_f

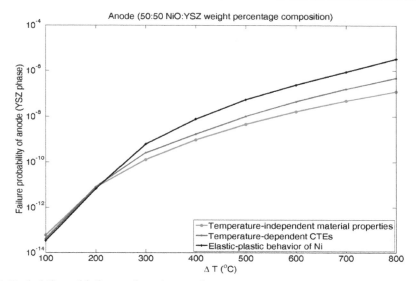

Fig. 10. Probability of failure values for anode

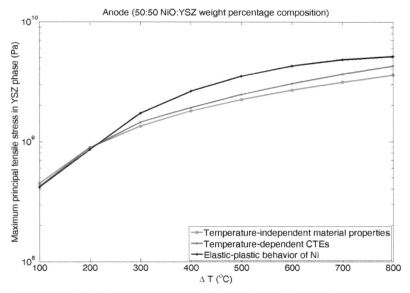

Fig. 11. Maximum principal tensile stress in the YSZ phase of the anode

values of the YSZ phase), as compared with the model that considers nonlinear elastic-plastic behavior of nickel, especially at high temperatures. This may be explained by referring to Figure 11, which shows that the maximum principal tensile stress (MPTS) in the YSZ phase of the anode increases with increasing ΔT values for all three cases, as expected. Figure 11 also shows that when the elastic-plastic behavior of Ni is taken into account, the

MPTS in the YSZ phase attains higher values than when linear elastic behavior is assumed, especially at high temperatures. This can be explained as follows: when the Ni phase enters the nonlinear (plastic) part of its stress-strain curve at higher temperatures (and hence higher strains), lower stresses are induced in the Ni phase than if its stress-strain curve had been purely linear elastic with the same value of Young's modulus. Thus, when the Ni phase starts showing nonlinear behavior, a higher proportion of the temperature-induced stresses are redistributed into the YSZ phase, resulting in higher MPTS values in the YSZ phase (and hence higher P_f values for the anode).

Figure 10 also shows that the case with temperature-dependent CTEs shows higher P_f values than the case with temperature-independent material properties at intermediate and high temperatures. Again, Figure 11 shows that with temperature-dependent CTE values, higher tensile stresses are induced in the YSZ phase of the anode than with temperature-independent material properties, especially at intermediate and high temperatures. This can be explained by referring to Figure 4, which shows that the CTEs of both Ni and YSZ increase with temperature. Since thermal stresses are proportional to CTE values, it can be expected that the case with temperature-dependent CTEs will show higher MPTS values (and hence higher P_f values) than the case with temperature-independent material properties, which uses constant (room-temperature) values of the CTEs.

4.2.2 Cathode

The probability of failure (P_f) values for the LSM and YSZ phases of the cathode were calculated and combined, as described above, at each ΔT value (100°C, 200°C, ..., 800°C) for both the cases described in Table 2. These values are plotted in Figure 12.

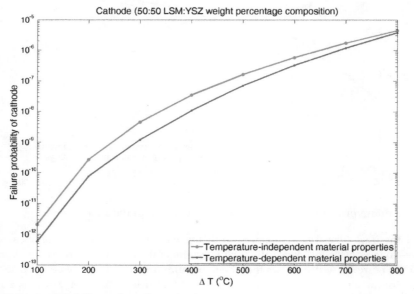

Fig. 12. Failure probability values for cathode

The P_f plot for the cathode shows that the probability of failure of the cathode increases with increasing ΔT values (and hence increasing stresses), for both temperature-independent and temperature-dependent material properties, as expected. Higher P_f values are obtained when temperature-independent material properties are considered. A physical explanation for this observation is suggested by the temperature variation of the Young's modulus of YSZ. For YSZ, E decreases from a value of 205 GPa at T = 20°C to a value of 147.5 GPa at T = 800°C, as shown in Figure 5. On the other hand, when temperature-independent material properties are considered, the Young's modulus of YSZ has a constant value of 205 GPa. Thus, because of the large decrease in the Young's modulus of YSZ with increasing temperature, lower stresses are induced in the cathode in the case with temperature-dependent material properties than in the case with temperature-independent material properties. This in turn leads to lower P_f values in the case with temperature-dependent material properties as compared with the case that considers temperature-independent material properties. This is confirmed by the MPTS plot for the cathode shown below (Figure 13), which compares the maximum principal tensile stress induced in the YSZ and LSM phases of the cathode for temperature-independent and temperature-dependent material properties.

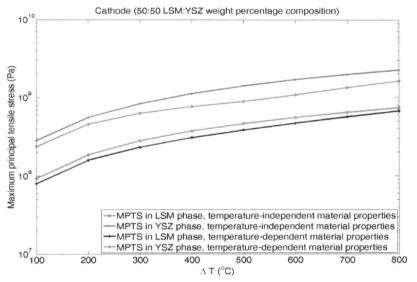

Fig. 13. Maximum principal tensile stress in LSM and YSZ phases of cathode

The plot above shows that the MPTS induced in the LSM phase for temperature-dependent material properties is lower than the MPTS in the LSM phase for temperature-independent material properties over the entire temperature range. Similarly, the MPTS induced in the YSZ phase for temperature-dependent material properties is lower than the MPTS induced in the YSZ phase for temperature-independent material properties over the entire temperature range. This implies that the cathode P_f values, which are calculated on the basis of the positive (tensile) principal stresses in the LSM and YSZ phases, will be higher for the temperature-independent material properties case than for the temperature-dependent material properties case, as is indeed observed.

5. Conclusions

Three-dimensional FE models of SOFC anode and cathode microstructures were constructed from a stack of two-dimensional SEM images of actual cross-sections of anode and cathode microstructures. The models were subjected to spatially uniform predefined temperature fields of increasing magnitude and the resulting distribution of stresses was obtained using FEA. The obtained stresses were subjected to Weibull analyses to determine the failure probability of the anode and cathode as a function of temperature. *The novelties of this work include FE analysis of the mechanical response of microstructure-based anode and cathode models to temperature loads, consideration of temperature-dependent material properties of the anode and cathode materials, and consideration of nonlinear elastic-plastic behavior of the nickel phase of the Ni-YSZ anode.* The Weibull analyses showed that the linear elastic material models underestimate the failure probability of the anode at high temperatures; hence, it is important to consider the nonlinear behavior of the nickel phase of the Ni-YSZ anode. Also, it was found that consideration of temperature-independent material properties of the cathode materials results in higher failure probability values than those obtained with temperature-dependent material properties.

6. Acknowledgements

We acknowledge the financial support for this work from the National Science Foundation under the Faculty Early Career Development (CAREER) Grant CMMI-0546225 (Material Design & Surface Engineering Program). We also acknowledge the technical support from Dr. Scott Barnett at the Department of Materials Science and Engineering at Northwestern University who generously provided a series of 2-D SEM images of anode and cathode microstructures.

7. References

Anandakumar, G., Li, N., Verma, A., Singh, P., & Kim, J.-H. (2010). Thermal stress and probability of failure analyses of functionally graded solid oxide fuel cells. *Journal of Power Sources*, Vol. 195, (2010), pp. (6659-6670).

Atkinson, A., & Selcuk, A. (2000). Mechanical behavior of ceramic oxygen ion-conducting membranes. *Solid State Ionics*, Vol. 134, (2000), pp. (59-66).

Ebrahimi, F., Bourne, G., Kelly, M., & Matthews, T. (1999). Mechanical properties of nanocrystalline nickel produced by electrodeposition. *NanoStructured Materials*, Vol. 11, No. 3, (1999), pp. (343-350).

Giraud, S., & Canel, J. (2008). Young's modulus of some SOFCs materials as a function of temperature. *Journal of the European Ceramic Society*, Vol. 28, (2008), pp. (77-83).

Johnson, J., & Qu, J. (2008). Effective modulus and coefficient of thermal expansion of Ni-YSZ porous cermets. *Journal of Power Sources*, Vol. 181, (2008), pp. (85-92).

Kramer, J., Mastronarde, D., & McIntosh, J. (1996). Computer visualization of three-dimensional image data using IMOD. *Journal of Structural Biology*, Vol. 116, (1996), pp. (71-76).

Laurencin, J., Delette, G., Lefebvre-Joud, F., & Dupeux, M. (2008). A numerical tool to estimate SOFC mechanical degradation: case of the planar cell configuration. *Journal of the European Ceramic Society*, Vol. 28, (2008), pp. (1857-1869).

Meyers, M., & Chawla, K. (1999). *Mechanical Behavior of Materials*, Prentice Hall, ISBN 0132628171, Upper Saddle River, New Jersey.

Nakajo, A., Stiller, C., Harkegard, G., & Bolland, O. (2006). Modeling of thermal stresses and probability of survival of tubular SOFC. *Journal of Power Sources*, Vol. 158, (2006), pp. (287-294).

Pihlatie, M., Kaiser, A., & Mogensen, M. (2009). Mechanical properties of NiO/Ni-YSZ composites depending on temperature, porosity and redox cycling. *Journal of the European Ceramic Society*, Vol. 29, (2009), pp. (1657-1664).

Pitakthapanaphong, S., & Busso, E. (2005). Finite element analysis of the fracture behaviour of multi-layered systems used in solid oxide fuel cell applications. *Modelling and Simulation in Materials Science and Engineering*, Vol. 13, (2005), pp. (531-540).

Selcuk, A., & Atkinson, A. (1997). Elastic properties of ceramic oxides used in solid oxide fuel cells (SOFC). *Journal of the European Ceramic Society*, Vol. 17, (1997), pp. (1523-1532).

Selcuk, A., & Atkinson, A. (2000). Strength and toughness of tape-cast yttria-stabilized zirconia. *Journal of the American Ceramic Society*, Vol. 83, No. 8, (2000), pp. (2029-2035).

Singhal, S., & Kendall, K. (Eds.). (2003). *High Temperature Solid Oxide Fuel Cells: Fundamentals, Design and Applications*, Elsevier, ISBN 1856173879, Oxford.

Toftegaard, H., Sorensen, B., Linderoth, S., Lundberg, M., & Feih, S. (2009). Effects of heat-treatments on the mechanical strength of coated YSZ: an experimental assessment. *Journal of the American Ceramic Society*, Vol. 92, No. 11, (2009), pp. (2704-2712).

Toftegaard, H., & Sorensen, B. (2009). Effects of heat-treatments on the mechanical strength of coated YSZ: an experimental assessment. *Journal of the American Ceramic Society*, Vol. 92, No. 11, (2009), pp. (2704-2712).

Weibull, W. (1951). A statistical distribution function of wide applicability. *ASME Journal of Applied Mechanics*, (1951), pp. (293-297).

Wilson, J., Kobsiriphat, W., Mendoza, R., Chen, H.-Y., Hiller, J., Miller, D., Thornton, K., Voorhees, P., Adler, S., & Barnett, S. (2006). Three-dimensional reconstruction of a solid-oxide fuel-cell anode. *Nature Materials*, Vol. 5, (2006), pp. (541-544).

Wilson, J., & Barnett, S. (2008). Solid oxide fuel cell Ni-YSZ anodes: effect of composition on microstructure and performance. *Electrochemical and Solid-State Letters*, Vol. 11, No. 10, (2008), pp. (B181-B185).

Wilson, J., Duong, A., Gameiro, M., Chen, H.-Y., Thornton, K., Mumm, D., & Barnett, S. (2009). Quantitative three-dimensional microstructure of a solid oxide fuel cell cathode. *Electrochemistry Communications*, Vol. 11, (2009), pp. (1052-1056).

Xiao, C., Mirshams, R., Whang, S., & Yin, W. (2001). Tensile behavior and fracture in nickel and carbon doped nanocrystalline nickel. *Materials Science and Engineering A*, Vol. 301, (2001), pp. (35-43).

Zhang, T., Zhu, Q., Huang, W., Xie, Z., & Xin, X. (2008). Stress field and failure probability analysis for the single cell of planar solid oxide fuel cells. *Journal of Power Sources*, Vol. 182, (2008), pp. (540-545).

3

Transversality Condition in Continuum Mechanics

Jianlin Liu

Department of Engineering Mechanics, China University of Petroleum,
China

1. Introduction

Nature creates all kinds of miraculous similar phenomena in the real world. For example, the spiral morphologies exist in nebula, sunflower seed array, grapevine, and DNA. There are also a lot of similarities in physical theories and principles, such as the analogy between a harmonic vibration system and an RLC oscillation circuit, between a membrane and a sand-heap in elasticity and plasticity, and between fluid mechanics and electricity or magnetism. The great scientist Maxwell pointed out that the form of the capillary surface is identical with that of the elastic curve, which was later tested by the experiment of Clanet and Quere (2002), and then was analyzed by Liu in detail (2009). Exploring these similarities and analogies can help us understand the underlining secret of nature, and pave the way to incorporate several similar phenomena into a unified analysis frame.

For this study, we mainly focus on the similarity in the adhesion of materials and devices at micro and nano scales, which may be caused by van der Waals force, Casimir force, capillary force or other interaction forces. Among others, the adhesion of a slender structure as micro-beam or carbon nanotube (CNT) is of great value for both theoretical and practical aspects. In these systems, due to considerable surface to volume ratio in low-dimensional micro/nano-systems, surface tension or interfacial energy will dominate over the volume force as their dimensions shrink to micro/nano-meters, which presents a lot of novel behaviors distinct with those of the macroscopic systems (Poncharal, et al., 1999). The typical phenomenon is stiction of the micro-beams, such as the micro/nano-wires and micro/nano-belts which are widely used as building blocks of micro-sensors, resonators, probes, transistors and actuators in M/NMES (micro/nano-electro-mechanical systems). In micro-contact printing technology, adhesion associated with van der Waals force leads to stamp deformation (Hui, et al., 2002), and the micro-machined MEMS structures will spontaneously adhere on the substrate under the influence of solid surface energy or liquid surface tension (Zhao, et al., 2003). This failure due to stiction has become a major limitation to push the better application of these novel engineering devices, and the problem has been highlighted as a hot topic in the past decades. The main reason of stiction is that in the small spacings, the slender structures with high compliance are easily brought into contact with the substrate of strong surface energy.

Another related issue is the self-collapse of a single wall carbon nanotube (SWCNT), in which process its initially circular cross-section will jump to a flat ribbon like shape. The reason lies in that CNTs capture the characteristic of hollow cylindrical structures, which render them susceptible to lateral deformation. In reality, this morphology was first observed and explored by transmission electron microscopy (TEM) (Ruoff, et al., 1993; Chopra, et al., 1995) and then by AFM (Martel, et al., 1998; Yu, et al., 2000). To date, much effort has been devoted towards understanding the mechanism of CNT collapse. Gao *et al.* (1998) used molecular dynamic (MD) simulations to discover that there are two possible configurations for a CNT in equilibrium state, i.e. the circular one and the collapsed one. For nanotubes with radius in the range of $R < R_{min}$, only the circular configuration exits. When the radius satisfies $R_{min} < R < R_{max}$, both of the shapes exist, and the collapsed tube is in a metastable state. For the radius $R > R_{max}$, the collapsed configuration is energetically favorable and thermodynamically stable. Their results also exhibited that the critical radii are insensitive to the chirality of the tube, and the values of the critical radii are $R_{min} \approx 1.1$ nm and $R_{max} \approx 3$ nm. Subsequently, Pantano *et al.* (2004) adopted a continuum approach and finite element method to investigate the morphology of the collapsed CNTs. In succession, Tang *et al.* studied the collapse of nanotubes using an inextensible elastica model. In use of phase plane analysis, they showed that CNTs can take collapsed configurations of different orders (Tang and Glassmaker, 2010). Recently, they investigated the energetics of self-collapse of a single CNT by using the continuum mechanics method, and calculated the critical radii $R_{min} \approx 0.699$ nm and $R_{max} \approx 0.976$ nm (Tang, et al., 2005b). This significant difference from the result of Gao *et al.* is because a distinct force field and physical parameters were selected, which greatly affect the results of analytical solutions and molecular simulations. However, Liu *et al.* (2004) performed simulations on the formation of fully collapsed SWCNTs with the atomic scale finite element method, and proposed that for armchair SWCNTs, collapse occurs for the critical radius (for n=30) is $R_{max} \approx 2.06$ nm, which is also different from the aforementioned results. In fact, this value of critical radius of 2.06 nm was verified by TEM observations, which demonstrated that there exists a collapsed SWCNT of 2.5 nm in radius (Xiao, et al., 2007). In spite of the above cited studies on the collapse of CNTs, there are hitherto no analytical solution for the collapse problem of CNTs, which involves large deformation and strong geometric nonlinearity.

Although belonging to different phenomena, we strongly stress that, the adhesion of micro-beams and collapse of SWCNT can be actually incorporated into a unified analysis frame. Based upon this frame, we can easily calculate the parameters of adhesion for different systems. The outline of this article is organized as follows. In Section 2, we established the formulations of energy functional, and derived the governing equation and transversality condition in consideration of the moving boundary. In Section 3, in use of the constructed frame, we obtained the critical adhesion length of two micro-beams stuck by a thin liquid film, and the deflections of the beams. In Section 4, we calculated the critical radii and collapsed shapes of SWCNTs via the classical elastica solution, which was derived from the energy functional.

2. Energy functional and transversality condition

We first provide the analysis of the scaling law of a system with different energy originations. The typical length of the slender structure is denoted as L_c, then the interfacial or surface energy $U_S \propto L_c$, the elastic strain energy $U_E \propto L_c^2$, and the potential energy due to gravity $U_G \propto L_c^3$ (Roman and Bico, 2010). As the dimension of a macroscopic structure shrinks to nanometers, the effect of surface energy will become significant and should be taken into account. In this study, we assume that the gravity effect is negligible, for the interplay between the surface energy and elasticity is predominant.

Let us consider a generalized elastic system denoted by a continuous and smooth curve, where part of the curve is adhered by some special interfacial forces. The position of an arbitrary point in the curve is schematized by the arc length s, the total length of the curve is L, and the segment length dealing with the elastic deformation is a. The kernel problem is how to determine the unknown length a in the equilibrium state according to the principle of least potential energy.

The functional of the total potential energy about the system is normally written as:

$$\Pi\left[y(x,a)\right] = \int_0^a U_E dx - \int_a^L W_a dx , \tag{1}$$

where the first term in the right side of Eq. (1) is strain energy, $U_E = F\left[x, y(x,a), y'(x,a), y''(x,a)\right]$, and the symbols $( )' = \dfrac{\partial( )}{\partial x}$, $( )'' = \dfrac{\partial^2( )}{\partial x^2}$.

The second term in the right side of Eq. (1) is named as the work of adhesion between two surfaces, which is normally expressed as (Tang, et al., 2005a)

$$W_a = \left(\gamma_1 + \gamma_2 - \gamma_{12}\right)b , \tag{2}$$

where b is the contact width out of the curve, γ_1 and γ_2 are the surface energies of the two different phases, and γ_{12} the interfacial energy. In the conventional definition, the work of adhesion is actually the work per unit area necessary to create two new surfaces from a unit area of an adhered interface, which is a positive constant for any two homogeneous materials binding at an interface at a fixed temperature. For the two phases are of the same material, the work of adhesion degenerates to the cohesive work

$$W_c = 2\gamma_1 b . \tag{3}$$

At micro and nano scales, the cohesive work is normally termed as the binding energy E_B. Furthermore, for the interface consisting of a thin liquid film, the expression of the cohesive work is (Bico, et al., 2004)

$$W_c = 2\left(\gamma_{SV} - \gamma_{SL}\right)b = 2\gamma\cos\theta_Y b , \tag{4}$$

where γ_{SV}, γ_{SL}, γ are the interfacial tensions of solid/vapor, solid/liquid and liquid/vapor interfaces, respectively, with θ_Y being the Young's contact angle of the liquid. In the above derivation, the Young's equation is employed.

We should mention that, the energy functional of Eq. (1) actually includes two variables, namely, the function y and a, because a is yet an unknown when solving the governing equation. This results to an intractable problem, for the undetermined variable a causes the bound movement of the system, which should be considered as a movable boundary condition problem during variation process. Therefore, in use of the principle of least potential energy, one obtains the following variational result

$$\delta \Pi \left[y(x,a) \right] = \delta \Pi_1 + \delta \Pi_2 = 0 , \tag{5}$$

where

$$\delta \Pi_1 = \int_0^a \left(F_y \delta y + F_{y'} \delta y' + F_{y''} \delta y'' \right) dx$$

$$= \left[F_{y'} \delta y + F_{y''} \delta y' - \frac{\partial}{\partial x} F_{y''} \delta y \right]_0^a + \int_0^a \left(F_y - \frac{\partial}{\partial x} F_{y'} + \frac{\partial^2}{\partial x^2} F_{y''} \right) \delta y dx . \tag{6}$$

For the other portion of variation, in consideration of the moving boundary at $s=a$, we then focus on the transversality condition in mathematical meaning, and show how to derive its expression. We first revisit the derivative definition about an integration including a parameter α. Let

$$\phi(\alpha) = \int_{a(\alpha)}^{b(\alpha)} F(x,\alpha) dx , \tag{7}$$

and then we have its derivative

$$\phi'(\alpha) = \int_{a(\alpha)}^{b(\alpha)} F_\alpha (x,\alpha) dx + F\left[b(\alpha),\alpha \right] b'(\alpha) - F\left[a(\alpha),\alpha \right] a'(\alpha) . \tag{8}$$

Similarly, considering the moving boundary at $s=a$, we can obtain the second part of variation on the functional in Eq. (1), that is

$$\delta \Pi_2 = \left[F - y' F_{y'} - y'' F_{y''} + y' \frac{dF_{y''}}{dx} + W \right]_{x=a} \delta a . \tag{9}$$

In the above derivations, the partial derivative of the function F is designated as $F_{(\ )} = \frac{\partial F}{\partial (\ )}$.
The forced or fixed boundary conditions are normally prescribed as

$$y(0)=y_0, \ y'(0) = y'_0 \ ; \ y(a)=y_a, \ y'(a) = y'_a . \tag{10}$$

Inserting Eq. (6) and (9) into (5), and according to the arbitrariness of the variation, one can get the governing differential equation, i.e. the Euler-Poisson equation:

$$F_y - \frac{\partial}{\partial x} F_{y'} + \frac{\partial^2}{\partial x^2} F_{y''} = 0 . \tag{11}$$

Besides the above equation, the arbitrariness of variation about a leads to

$$W_a = \left[y' F_{y'} + y'' F_{y''} - y' \frac{dF_{y''}}{dx} - F \right]_{x=a} . \tag{12}$$

In fact, the concept of energy release rate or J integral was also adopted to investigate the problem of moving boundary (Tang et al., 2005a, 2005b). However, the exact solution of the energy release rate is often impossible to acquire, as a result, one can avoid this way and apply the transversality condition to determine the unknown a. This thought paves a new way to solve this kind of problems dealing with movable boundary. In addition, the transversality condition is essential to calculate the contact angle of a droplet, the morphology of a cell, and the peeling of a CNT from the substrate (Oyharcalbal and Frisch, 2005; Bormasshenko and Whyman, 2008; Seifert, 1990). We will then put to use this method to analyze some practical topics, and the adhesion of two micro-beams and the collapsed shape of a single wall carbon nanotube (SWCNT) are selected as study cases.

3. Adhesion of two micro-beams

Let us consider two identical micro-beams with the same Young's modulus E, and moment of inertia on the cross section I, which are stuck together by the interfacial energy W_a (or work of adhesion) due to a thin liquid film. The cross section of the beam is a rectangle, with the width b, and thickness e, then $I = be^3/12$. Refer to a Cartesian coordinate system $(o\text{-}xy)$. As shown in Fig. 1, the distance between the ends of the two beams is d, the detached segment length is L_{dry}, the adhered part is L_{wet}, and the total length of the beam is L.

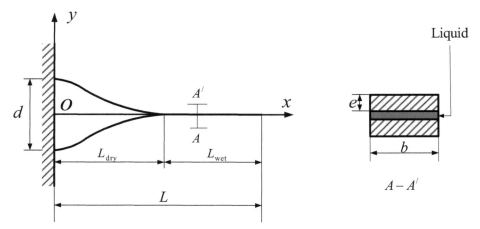

Fig. 1. Capillary adhesion of two beams with rectangular cross-sections.

According to Eqs. (1) and (4), the total potential energy of the two beam system can be easily expressed as (Bico, et al., 2004; Liu and Feng, 2007a)

$$\Pi = EI \int_0^{L_{dry}} w''^2 dx - 2\gamma \cos\theta_Y \left(L - L_{dry} \right). \tag{13}$$

In use of Eqs. (11) and (12), one can deduce the governing equation

$$w^{(4)} = 0, \tag{14}$$

and the transversality condition at the moving bound

$$2\gamma \cos\theta_Y b = EI w'' \left(L_{dry} \right)^2. \tag{15}$$

The fixed boundary conditions of a single beam are specified as

$$w(0)=d/2, \; w'(0)=0 \; ; w(L_{dry})=0, \; w'\left(L_{dry}\right)=0, \tag{16}$$

Then the corresponding solution of Eq. (14) is written as

$$w = \frac{d}{L_{dry}^3}x^3 - \frac{3d}{2L_{dry}^2}x^2 + \frac{d}{2}. \tag{17}$$

The combination of Eq. (15) and (17) yields

$$L_{dry} = \sqrt[4]{\frac{3Ee^3d^2}{8\gamma\cos\theta_Y}}. \tag{18}$$

If the total length of the beam $L < L_{dry}$ or $\tilde{L}_{dry} = L_{dry}/L > 1$, the adhesion energy induced by the introduction of a liquid film between the two beams is insufficient to provide the strain energy of deformation, and therefore, the two beams will not adhere. On the contrary, if $\tilde{L}_{dry} < 1$, the surface energy is larger than the strain energy, and then the adhesion of the two beams is possible.

Equation (18) also requires that the contact angle θ_Y must satisfy $0 \le \theta_Y < \pi/2$, that is, the beams must be hydrophilic. In other words, capillary adhesion cannot happen between two hydrophobic hairs. The critical length L_{dry} for capillary adhesion of two beams increases with the decreasing of the surface tension and the Young's contact angle θ_Y in the range of $0 \le \theta_Y < \pi/2$, and with the increasing of the end distance d of the two beams. In addition, the deflection diagrams of the two beams can be determined easily from Eq. (17) and are plotted in Fig. 2 for several representative values of the interbeam spacing, $d=0.5, 1.0$ and 2.0 mm, where we take the following parameters: $EI/b= 5.1\times10^{-4}$ N·m , $\theta_Y = 10°$ and $\gamma = 72\times10^{-3}$ N/m.

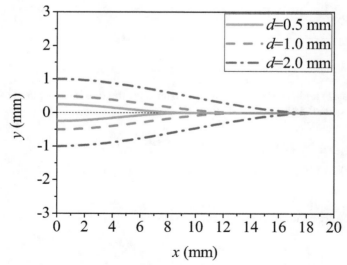

Fig. 2. Deflections of two adhered micro-beams by the liquid film.

4. Collapsed morphology of a SWCNT

We then put to use our analysis frame to examine the collapsed morphology of a SWCNT, which is initially circular with a radius R and an axial length L. The current configuration incorporates a flat contact zone in the middle part and two non-contact regions at the ends, as shown in Fig. 3. In essence, this configuration is stabilized by the van der Waals interplay between the upper and lower portion of the CNT walls, primarily within the horizontal contact zone, because the van der Waals force decays rapidly in the non-contact areas. As a reasonable simplification, the van der Waals force between the upper and lower portion of the CNT walls in the non-contact domain is ignored in our calculation. Normally, the van der Waals force between two carbon atoms is repulsive at a very close range, so the CNT wall contact is defined by an equilibrium separation d_0 between the flat regions. The distance between the flat contact zone and the extreme point of the CNT is denoted as b. From the experimental picture, we can see that the collapsed shape of CNT is symmetric, which was also verified by the molecular simulations (Tang, et al. 2005b).

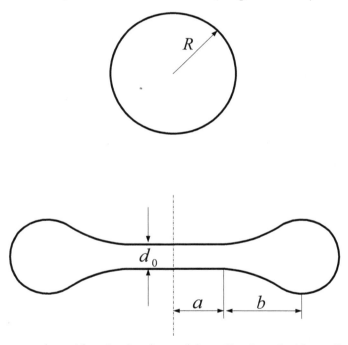

Fig. 3. Carbon nanotubes with a circular shape of the radius R, and with a collapsed shape. The semi-width of the flat contact zone of the collapsed CNT is a, and the separation distance is d_0.

Due to the symmetry and smoothness of this configuration, a quarter of the structure is selected and then modeled as a plate or an elastica with two clamped ends, as schematized in Fig. 4. This famous elastica theory, which can be traced back to the historic contribution of Euler (Love, 1906), has been successfully used to solve some finite deformation problems of slender structures (Bishopp, 1945; Liu and Feng, 2007a; Glassmaker and Hui, 2004).

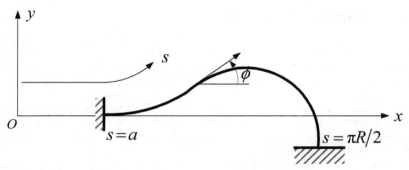

Fig. 4. Elastica model for a quarter of the CNT, with the arc length s and slope angle ϕ ranging from 0 to $-90°$.

In fact, there is another underling assumption that, the deformation along the axis of the nanotube is uniform, which has already been verified by experiments and MD simulations (Ruoff, et al. 1993; Chopra, et al., 1995; Pantano, 2004; Tang, et al., 2005b). As a result, we select the cross section representing the whole tube, and model the thin wall as a curvilinear abscissa. The total in-plane length of the elastica is thus $\pi R / 2$, and the adhered segment is a. Refer to a Cartesian coordinate system (o-xy). Besides the Euler coordinate x and y, the arc length s, being a Lagrange coordinate, is also employed as a variable in our analysis. The slope angle of the elastica at an arbitrary point is ϕ, which continuously changes from $0°$ at its left end to $-90°$ at the right end. The bending stiffness of the elastica is EI, where I is the inertial moment of the cross section, and $E = \overline{E}/(1-\upsilon^2)$, with $\overline{E}$ being the Young's modulus and υ the Poisson's ratio of the material.

According to the elastica model of Fig. 4, the fixed displacement boundary conditions of the system are specified as

$$\phi(a) = 0, \ \phi\left(\frac{\pi R}{2}\right) = -\frac{\pi}{2}, \ \phi(\pi R - a) = -\pi ; \ y(a) = 0, \ y\left(\frac{\pi R}{2}\right) = -\frac{d_0}{2}. \tag{19}$$

The additional geometric conditions of the elastica are

$$\dot{x} = \cos\phi, \ \dot{y} = \sin\phi, \tag{20}$$

where the dot above a parameter stands for its derivative with respect to the arc length s. In consideration of the symmetry of this configuration, the total potential energy functional of the system can be written as

$$\Pi = 2L\int_a^{\pi R/2}\left[EI\dot{\phi}^2 + \lambda_1(\dot{x} - \cos\phi) + \lambda_2(\dot{y} - \sin\phi)\right]ds - 2W_c aL, \tag{21}$$

where λ_1 and λ_2 are two Lagrange multipliers, enforcing the additionally geometrical relations of Eq. (20).

To deduce the expression of the cohesive work, we choose the van der Waals potential as (Tang, et al., 2005a)

$$V(r) = \varepsilon \left[3 \left(\frac{r^*}{r} \right)^6 - 2 \left(\frac{r^*}{r} \right)^9 \right], \tag{22}$$

where ε =0.064 kcal/mol, r^* =0.401 nm, and r is the distance between two atoms. In this case, the cohesive work is the work per unit area to separate the two parallel graphite sheets from d to ∞, which can be calculated via integration of Eq. (22):

$$W(d) = \pi \rho^2 \varepsilon \left[\frac{3(r^*)^6}{2d^4} - \frac{4(r^*)^9}{7d^7} \right], \tag{23}$$

where ρ is the number of carbon atoms per unit area, with the value of $\rho \approx 0.004$ nm^{-2}. In the experiment (Ruoff, et al. 1993), the equilibrium distance is measured as d_0=0.338 nm, and the corresponding cohesive work is calculated as W_c=0.388627 J/m^2 according to Eq. (23).

Taking variation of the potential energy functional of Eq. (21), and in use of the principle of least potential energy, one has

$$\delta \Pi = \delta \Pi_1 + \delta \Pi_2 = 0, \tag{24}$$

where

$$\delta \Pi_1 = 2L \int_a^{\pi R/2} \left[2EI \dot{\phi} \delta \dot{\phi} + \delta \lambda_1 \left(\dot{x} - \cos \phi \right) + \delta \lambda_2 \left(\dot{y} - \sin \phi \right) \right.$$

$$\left. + \lambda_1 \delta \dot{x} + \lambda_1 \sin \phi \delta \phi + \lambda_2 \delta \dot{y} - \lambda_2 \cos \phi \delta \phi \right] ds$$

$$= 2L \int_a^{\pi R/2} \left[-2EI \ddot{\phi} \delta \phi + \delta \lambda_1 \left(\dot{x} - \cos \phi \right) + \delta \lambda_2 \left(\dot{y} - \sin \phi \right) \right.$$

$$\left. + \lambda_1 \sin \phi \delta \phi - \lambda_2 \cos \phi \delta \phi \right] ds \ + 2L \left[2EI \dot{\phi} \delta \phi + \lambda_1 \delta x + \lambda_2 \delta y \right]_a^{\pi R/2} \tag{25}$$

$$= 2L \int_a^{\pi R/2} \left[\delta \lambda_1 \left(\dot{x} - \cos \phi \right) + \delta \lambda_2 \left(\dot{y} - \sin \phi \right) \right.$$

$$\left. - \left(2EI \ddot{\phi} - \lambda_1 \sin \phi + \lambda_2 \cos \phi \right) \delta \phi \right] ds \ - 2L \lambda_1 \delta a.$$

In the above derivations, Eqs. (19) and (20) have been adopted.

Considering the moving boundary, the second part of the variation is expressed as

$$\delta \Pi_2 = -2L \left[EI \dot{\phi}^2 + \lambda_1 \left(\dot{x} - \cos \phi \right) + \lambda_2 \left(\dot{y} - \sin \phi \right) - 2\dot{\phi} \left(EI \dot{\phi} \right) - \dot{x} \lambda_1 - \dot{y} \lambda_2 \right]_{s=a} \delta a$$

$$- 2W_c L \delta a \tag{26}$$

$$= 2L \left[EI \dot{\phi} (a)^2 - W_c + \lambda_1 \right] \delta a.$$

Noticing the arbitrariness of the variation, one obtains the following governing equation

$$2EI\ddot{\phi} - \lambda_1 \sin\phi + \lambda_2 \cos\phi = 0 \,, \tag{27}$$

where the Lagrange multipliers λ_1 and λ_2 can be easily identified as the horizontal and vertical internal forces in the elastica. The governing equation in Eq. (27) conforms to that derived by Tang et al. (Tang and Glassmaker, 2010), who adopted the method of force equilibrium.

Combination of Eqs. (24) and (26) leads to the additional boundary condition at the moving point

$$EI\dot{\phi}(a)^2 - W_c = 0 \,, \tag{28}$$

which is termed as the transversality condition. This additional condition represents the balance of the elastic strain energy and the van der Waals potential energy. It is worthy of being mentioned that the movable boundary condition in Eq. (28) can also be deduced via the concept of J integral in fracture mechanics, as described by Glassmaker and Hui (2004) in their analysis of silicon–germanium nanotube formation.

Multiplying $\dot{\phi}$ to both sides of Eq. (27) and by integration, one has

$$EI\dot{\phi}^2 = D - \lambda_1 \cos\phi - \lambda_2 \sin\phi \,, \tag{29}$$

where D is an integration constant. Making use of Eq. (19), one obtains

$$EI\dot{\phi}(a)^2 = D \,. \tag{30}$$

It is noticed that the symmetry of the configuration verifies the relation of $\dot{\phi}(a)^2 = \dot{\phi}(\pi R - a)^2 = D$.

Inserting Eq. (28) into (30), one has

$$D = W_c. \tag{31}$$

Introducing the parameter C and $\alpha = \sqrt{\lambda_2/(2EI)}$, we can obtain $\alpha^2 C = D/(2EI)$. Thus the governing equation (29) and transversality condition (30) are respectively transformed into

$$\frac{1}{2}\dot{\phi}^2 = \alpha^2 (C - \sin\phi) \,, \tag{32}$$

$$\frac{1}{2}\dot{\phi}(a)^2 = \alpha^2 C \,. \tag{33}$$

The combination of Eqs. (30) and (33) yields

$$\alpha = \frac{1}{L_{ec}\sqrt{2C}} \,. \tag{34}$$

Here, we have defined a new characteristic length, i.e. the elasto-cohesive length L_{ec} = $\sqrt{EI/W_c}$, which is different from the elasto-capillary length L_{EC} named by Bico et al. (2004) and Roman (2010). For a slender structure adhered by a liquid film, the elasto-cohesive length $L_{ec} = \dfrac{\sqrt{2}}{2} L_{EC}$ when $\theta_Y = 0$. It is seen that in this case, Eqs. (30) and (34) are consistent with the results in the reference (Bico, et al., 2004).

Note that the arc increment ds is always positive and the increasing of the slope angle is not monotonic, and Eq. (32) is simplified as

$$\alpha ds = \frac{|d\phi|}{\sqrt{2(C-\sin\phi)}} . \tag{35}$$

For convenience of integration, the variable ϕ should be replaced with another variable θ. These two variables are related by

$$(1+C)\sin^2\theta = 2k^2 \sin^2\theta = 1 + \sin\phi \quad (0 \le \theta \le \pi, \ k > 0), \tag{36}$$

and

$$\sqrt{2(C-\sin\phi)} = 2k|\cos\theta|, \tag{37}$$

$$|d\phi| = \frac{2k|\cos\theta|}{\sqrt{1-k^2\sin^2\theta}} d\theta, \tag{38}$$

$$\frac{|d\phi|}{\sqrt{2(C-\sin\phi)}} = \frac{d\theta}{\sqrt{1-k^2\sin^2\theta}} . \tag{39}$$

Substituting Eqs. (34–39) into the prescribed displacement boundary condition in Eq. (19) leads to

$$\alpha y \left(\frac{\pi R}{2} \right) = \int_0^{-\frac{\pi}{2}} \frac{\sin\phi |d\phi|}{\sqrt{2(C-\sin\phi)}}$$

$$= F(k,\pi) - F(k,\theta_0) - 2\big[E(k,\pi) - E(k,\theta_0) \big] = -\frac{d_0}{2\sqrt{4k^2 - 2L_{ec}}}, \tag{40}$$

where $\sin\theta_0 = 1/(\sqrt{2}k)$, $F(k,\theta)$ and $E(k,\theta)$ are the elliptic integrals of the first and second kinds, which are respectively defined as

$$F(k,\theta) = \int_0^\theta \frac{1}{\sqrt{1-k^2\sin^2\theta}} d\theta, \tag{41}$$

$$E(k,\theta) = \int_0^\theta \sqrt{1-k^2\sin^2\theta}\, d\theta .$$

To close this mathematical formulation of the above problem, we then complement the inextensible condition of the elastic rod (Bishopp and Drucker, 1945), which is written as

$$\alpha\left(\frac{\pi R}{2}-a\right)=\alpha\int_{a}^{\pi R/2}ds$$

$$=\int_{0}^{-\frac{\pi}{2}}\frac{|d\phi|}{\sqrt{2(C-\sin\phi)}}=F(k,\pi)-F(k,\theta_{0}).$$

(42)

From Eq. (40), one can solve the corresponding values of $k=0.82$ for the given value of d_0. Then the substitution of Eq. (34) into (42) yields

$$\frac{a}{\pi R}=\frac{1}{2}-\sqrt{4k^{2}-2}\left[F(k,\pi)-F(k,\theta_{0})\right]\frac{L_{ec}}{\pi R}.$$

(43)

For a CNT with an initial radius $R=3$ nm, the analytical result of Eq. (43) is presented as 0.317854, which is nearly equal to the approximated solution given by Tang et al. (2005b) is $a/(\pi R)\approx(d_{0}-3.035L_{ec})/(\pi R)+0.5=0.30485$.

After the flat contact length a has been solved by Eq. (43), the deflection of the rod can be determined by

$$\alpha x=\alpha\int_{0}^{s}\cos\phi ds=\alpha a+\sqrt{4k^{2}-2}-2k\cos\theta,$$

(44)

$$\alpha y=\alpha\int_{0}^{s}\sin\phi ds=F(k,\theta)-F(k,\theta_{0})-2\left[E(k,\theta)-E(k,\theta_{0})\right].$$

(45)

The above displacements normalized by the elasto-cohesive length read

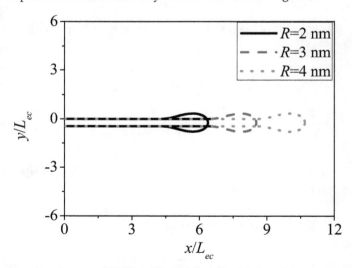

Fig. 5. Cross-section shapes of CNTs with initial radii of 2, 3 and 4 nm, respectively.

$$\begin{cases} \tilde{x} = \dfrac{\pi}{2}\tilde{R} - \sqrt{4k^2 - 2}\left[F(k,\pi) - F(k,\theta_0) - \sqrt{4k^2 - 2} + 2k\cos\theta \right], \\ \tilde{y} = \sqrt{4k^2 - 2}\left\{ F(k,\theta) - F(k,\theta_0) - 2\left[E(k,\theta) - E(k,\theta_0) \right] \right\} \end{cases} \qquad (46)$$

where $\tilde{x} = x/L_{ec}$, $\tilde{y} = y/L_{ec}$, and $\tilde{R} = R/L_{ec}$.

Finally, the morphologies of three representative collapsed CNTs with $R=2$, 3 and 4 nm are plotted in Fig. 5, where the parameters are the same as those in Fig. 3. It can be seen that, with increasing radius, the flat contact zone becomes larger, but the right ended shape does not change too much. Therefore, in the current calculation, the bending stiffness EI is selected as 2×10^{-19} N•m , and the corresponding elasto-cohesive length $L_{ec} = \sqrt{EI/W} = 0.72$ nm.

5. Conclusions

In this study, we demonstrated that a lot of problems dealing with the moving boundaries can be grouped into a unified frame, such as the adhesion of micro-beams and collapse of SWCNT. We first constructed the energy functional of the general system, then derived the governing equation and the transversality condition. We put this analysis method to solve the critical length and deflections of two micro-beams. Moreover, we derived the governing equation, i.e., the elastica model of the collapsed morphology for the SWCNT. Under the inextensible condition of the rod, the closed-form solutions for the flat contact segment, critical radii, and collapsed configuration were obtained in terms of elliptical integrals. It is clearly shown that our analytical solutions are in good agreement with the results of the references.

This analysis method paves a new way to examine nano-scaled mechanics by means of continuum mechanics. The presented results are also beneficial to design and fabricate new devices, micro-sensors and advanced materials in micro/nano scale, which casts a light on enhancing their mechanical, chemical, optical and electronic properties. Furthermore, this model can be generalized to investigate both a macroscopic sheet wrapped by a liquid film and a CNT self-folded by van der Waals forces, and can be adopted to analyze the crack or contact problems.

6. Acknowledgements

The project was supported by the National Natural Science Foundation of China (10802099), Doctoral Fund of Ministry of Education of China (200804251520), and Natural Science Foundation of Shandong Province (ZR2009AQ006). The author also acknowledges the support from the Brain Korea 21 program at Seoul National University.

7. References

Bico, J., Roman, B., Moulin, L., Boudaoud, A. (2004) Adhesion: elastocapillary coalescence in wet hair. *Nature*, 432, pp. 9.

Bishopp, K. E., Drucker, D. C. (1945) Large deflections of cantilever beams. *Quart. J. Appl. Math.*, 3, pp. 272–275.

Bormasshenko, E., Whyman, G. (2008) Variational approcach to wetting problems: calculation of a shape of sessile liquid drop deposited on a solid substrate in external field. *Chem. Phys. Lett.*, 463, pp. 103–105.

Chopra, N. G., Benedict, L. X., Crespi, V. H., Cohen, M. L., Louie, S. G., Zettl, A. (1995) Fully collapsed carbon nanotubes. *Nature*, 377, pp. 135-138.

Clanet, C., Quere, D. (2002) Onset of menisci. *J. Fluid Mech.*, 460, 2002 460, pp. 131-149.

Gao, G., Cagin, T., Goddard III, W. A. (1998) Energetics, structure, mechanical and vibrational properties of single-walled carbon nanotubes. *Nanotechnology*, 9, pp. 184-191.

Glassmaker, N. J., Hui, C. Y. (2004) Elastica solution for a nanotube formed by self-adhesion of a folded thin film. *J. Appl. Phys. 96*, pp. 3429-3444.

Hui, C. Y., Jagota, A., Lin, Y. Y., Kramer, E. J. (2002) Constraints on microcontact printing imposed by stamp deformation. *Langmuir*, 18, pp. 1394-1407.

Liu, B., Yu, M. F., Huang, Y. (2004) Role of lattice registry in the full collapse and twist formation of carbon nanotubes. *Phys. Rev. B*, 70, pp. 161402.

Liu, J. L. (2009) Analogies between a meniscus and a cantilever. *Chin. Phys. Lett.*, 26, pp. 116803.

Liu J L, Feng X Q. (2007) Hierarchical capillary adhesion of micro-cantilevers or hairs. *J. Phys. D: Appl. Phys.*, 40, pp. 5564-5570.

Liu, J. L., Feng, X. Q. (2007) Capillary adhesion of microbeams: finite deformation analyses. *Chin. Phys. Lett.*, 24, pp. 2349-2352.

Love, A.E.H. (1906) *A treatise on the mathematical theory of elasticity*, Second edition, Cambridge University Press, London.

Martel, R., Schmidt, T., Shea, H. R., Hertel, T., Avouris, P. (1998) Single- and multi-wall carbon nanotube field-effect transistors. *Appl. Phys. Lett.*, 73, pp. 2447-2449.

Oyharcalbal, X., Frisch, T. (2005) Peeling off an elastica from a smooth attractive substrate. *Phys. Rev. E*, 71, pp. 036611.

Pantano, A., Parks, D. M., Boyce, M. C., Buongiorno Nardelli, M. Mixed finite element-tight-binding electromechanical analysis of carbon nanotubes. *J. Appl. Phys.*, 96, pp. 6756-6760.

Poncharal, P., Wang, Z. L., Ugarte, D., de Heer, W. A. (1999) Electrostatic deflections and electromechanical resonances of Carbon Nanotubes. *Science*, 283, pp. 1513–1516.

Py, C., Reverdy, P., Doppler, L., Bico, J., Roman, B., Baroud, C. N. (2007) Capillary origami: spontaneous wrapping of a droplet with an elastic sheet. *Phys. Rev. Lett.*, 98, pp. 156103.

Roman, B., Bico, J. (2010) Elasto-capillarity: deforming an elastic structure with a liquid droplet. *J. Phys: Condens. Matter*, 22, pp. 493101.

Ruoff, R. S., Tersoff, J., Lorents, D. C., Subramoney, S., Chan, B. (1993) Radial deformation of carbon nanotubes by van der Waals forces. *Nature*, 364, pp. 514-516.

Seifert, U., Lipowsky, R. (1990) Adhesion of vesicles. *Phys. Rev. A*, 42, pp. 4768-4771.

Tang, T., Glassmaker, N. J. (2010) On the inextensible elastica model for the collapse of nanotubes. *Math. Mech. Solid*, 15, pp. 591-606.

Tang, T., Jagota, A., Hui, C. (2005) Adhesion between single-walled carbon nanotubes. *J. Appl. Phys.*, 97, pp. 074304.

Tang, T., Jagota, A., Hui, C. Y., Glassmaker, N. J. (2005) Collapse of single-walled carbon nanotubes. *J. Appl. Phys.*, 97, pp. 074310.

Xiao, J., Liu, B., Huang, Y., Zuo, J., Hwang, K. C., Yu, M. F. (2007) Collapse and stability of single- and multi-wall carbon nanotubes. *Nanotechnology*, 18, pp. 395703.

Yu, M. F., Dyer, M. J., Chen, J., Qian, D., Liu, W. K., Ruoff, R. S. (2001) Locked twist in multiwalled carbon-nanotube ribbons. *Phys. Rev. B*, 64, pp. 241403.

Yu, M. F., Kowalewski, T., Ruoff, R. S. (2000) Investigation of the radial deformability of individual carbon nanotubes under controlled indentation force. *Phys. Rev. Lett.*, 85, pp. 1456-1459.

Zhao, Y. P., Wang, L. S., Yu, T. X. (2003) Mechanics of adhesion in MEMS: a review. *J. Adhesion Sci. Technol.*, 17, pp. 519-546.

4

Incompressible Non-Newtonian Fluid Flows

Quoc-Hung Nguyen and Ngoc-Diep Nguyen
Mechanical Faculty, Ho Chi Minh University of Industry,
Vietnam

1. Introduction

A non-Newtonian fluid is a fluid whose flow properties differ in many ways from those of Newtonian fluids. Most commonly the viscosity of non-Newtonian fluids is not independent of shear rate or shear rate history. In practice, many fluid materials exhibits non-Newtonian fluid behavior such as: salt solutions, molten, ketchup, custard, toothpaste, starch suspensions, paint, blood, and shampoo etc. In a Newtonian fluid, the relation between the shear stress and the shear rate is linear, passing through the origin, the constant of proportionality being the coefficient of viscosity. In a non-Newtonian fluid, the relation between the shear stress and the shear rate is different, and can even be time-dependent. Therefore a constant coefficient of viscosity cannot be defined. In the previous parts of this book, the mechanics of Newtonian fluid have been mentioned. In this chapter, the common rheological models of non-Newtonian fluids are introduced and several approaches concerned with non-Newtonian fluid flows are considered. In addition, the solution of common non-Newtonian fluid flows in a circular pipe, annular and rectangular duct are presented.

2. Classification of non-Newtonian fluid

As above mentioned, a non-Newtonian fluid is one whose flow curve (shear stress versus shear rate) is nonlinear or does not pass through the origin, i.e. where the apparent viscosity, shear stress divided by shear rate, is not constant at a given temperature and pressure but is dependent on flow conditions such as flow geometry, shear rate, etc. and sometimes even on the kinematic history of the fluid element under consideration. Such materials may be conveniently grouped into three general classes:

1. fluids for which the rate of shear at any point is determined only by the value of the shear stress at that point at that instant; these fluids are variously known as 'time independent' , ' purely viscous' , 'inelastic' or 'generalized Newtonian fluids');
2. more complex fluids for which the relation between shear stress and shear rate depends, in addition, upon the duration of shearing and their kinematic history; they are called 'time-dependent fluids', and finally,
3. substances exhibiting characteristics of both ideal fluids and elastic solids and showing partial elastic recovery, after deformation; these are categorized as 'viscoplastic fluids'.

Among the three groups, the time independent Non-Newtonian fluids are the most popular and easiest to handle in analysis. In this chapter, only this group of Non-Newtonian fluids are considered.

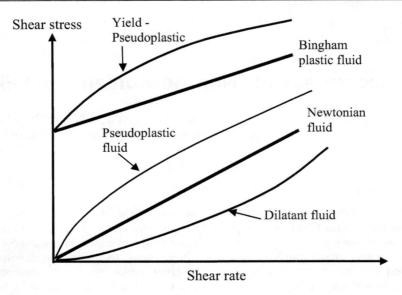

Fig. 1. Types of time-independent non-Newtonian fluid

In simple shear, the flow behaviour of this class of materials may be described by the following constitutive relation,

$$\tau_{yx} = f(\dot{\gamma}_{yx}) \tag{1}$$

This equation implies that the value of $\dot{\gamma}_{yx}$ at any point within the sheared fluid is determined only by the current value of shear stress at that point or vice versa. Depending upon the form of the function in equation (1), these fluids may be further subdivided into three types: shear-thinning or pseudoplastic, shear-thickening or dilatant and viscoplastic

2.1 Shear-thinning or pseudo-plastic fluids

The most common type of time-independent non-Newtonian fluid behaviour observed is Pseudo-plasticity or shear-thinning, characterized by an apparent viscosity which decreases with increasing shear rate. Both at very low and at very high shear rates, most shear-thinning polymer solutions and melts exhibit Newtonian behaviour, i.e., shear stress–shear rate plots become straight lines and on a linear scale will pass through origin. The resulting values of the apparent viscosity at very low and high shear rates are known as the zero shear viscosity, μ_0 , and the infinite shear viscosity, μ_∞, respectively. Thus, the apparent viscosity of a shear-thinning fluid decreases from μ_0 to μ_∞ with increasing shear rate. Many mathematical expressions of varying complexity and form have been proposed in the literature to model shear-thinning characteristics; some of these are straightforward attempts at curve fitting, giving empirical relationships for the shear stress (or apparent viscosity)–shear rate curves for example, while others have some theoretical basis in statistical mechanics – as an extension of the application of the kinetic theory to the liquid state or the theory of rate processes, etc. Only a selection of the more widely used viscosity models is given here; more complete descriptions of such models are available in many books (Bird *et al* ., 1987 ; Carreau *et al* ., 1997) and in a review paper (Bird, 1976).

i.   The power-law model

The relationship between shear stress and shear rate for this type of fluid can be mathematically expressed as follows:

$$\tau_{yx} = K(\dot{\gamma}_{yx})^n \tag{2}$$

So the apparent viscosity for the so-called power-law fluid is thus given by:

$$\mu = K(\dot{\gamma}_{yx})^{n-1} \tag{3}$$

For $n < 1$, the fluid exhibits shear-thinning properties
         $n = 1$, the fluid shows Newtonian behaviour
         $n > 1$, the fluid shows shear-thickening behaviour

In these equations, K and n are two empirical curve-fitting parameters and are known as the fluid consistency coefficient and the flow behaviour index respectively. For a shear thinning fluid, the index may have any value between 0 and 1. The smaller the value of n, the greater is the degree of shear-thinning. For a shear-thickening fluid, the index n will be greater than unity. When $n=1$, equations (3) becomes the constitutive equation of Newtonian fluid.

Although the power-law model offers the simplest representation of shear-thinning behaviour, it does have a number of limitations. Generally, it applies over only a limited range of shear rates and therefore the fitted values of K and n will depend on the range of shear rates considered. Furthermore, it does not predict the zero and infinite shear viscosities. Finally, it should be noted that the dimensions of the flow consistency coefficient, K, depend on the numerical value of n and therefore the K values must not be compared when the n values differ. On the other hand, the value of K can be viewed as the value of apparent viscosity at the shear rate of unity and will therefore depend on the time unit (e.g. second, minute or hour) employed. Despite these limitations, this is perhaps the most widely used model in the literature dealing with process engineering applications. Table 1 provides a compilation of the power-law constants (K and n) for a variety of substances.

ii.   The Carreau viscosity equation

When there are significant deviations from the power-law model at very high and very low shear rates, it is necessary to use a model which takes account of the limiting values of viscosities μ_0 and μ_∞ . Based on the molecular network considerations, Carreau (1972) put forward the following viscosity model.

$$\frac{\mu - \mu_\infty}{\mu_0 - \mu_\infty} = [1 + (\lambda \dot{\gamma}_{yx})^2]^{(n-1)/2} \tag{4}$$

where n (< 1) and λ are two curve-fitting parameters. This model can describe shear thinning behaviour over wide ranges of shear rates but only at the expense of the added complexity of four parameters. This model predicts Newtonian fluid behaviour $\mu = \mu_0$ when either $n = 1$ or $\lambda = 0$ or both.

System	Temperature (K)	n	M (Pa s^n)
Agro- and food-related products			
Ammonium alginate solution (3.37%)	297	0.5	13
Apple butter	–	0.15	200
Apple sauce	300	0.3–0.45	12–22
Apricot puree	300	0.3–0.4	5–20
Banana puree	293–315	0.33–0.5	4–10
Carrot puree	298	0.25	25
Chicken (minced)	296	0.10	900
Chocolate	303	0.5	0.7
Guava puree	296.5	0.5	40
Human blood	300	0.9	0.004
Mango pulp	300–340	0.3	3–10
Marshmallow cream	–	0.4	560
Mayonnaise	298	0.6	5–100
Papaya puree	300	0.5	10
Peach puree	300	0.38	1–5
Peanut butter	–	0.07	500
Pear puree	300	0.4–0.5	1–5
Plum puree	287	0.35	30–80
Tomato concentrate (5.8% solid)	305	0.6	0.22
Tomato ketch up	295	0.24	33
Tomato paste	–	0.5	15
Whipped desert toppings	–	0.12	400
Yoghurt	293	0.5–0.6	25
Polymer melts			
High density polyethylene (HDPE)	453–493	0.6	$3.75–6.2 \times 10^3$
High impact polystyrene	443–483	0.20	$3.5–7.5 \times 10^4$
Polystyrene	463–498	0.25	$1.5–4.5 \times 10^4$
Polypropylene	453–473	0.40	$4.5–7 \times 10^3$
Low density polyethylene (LDPE)	433–473	0.45	$4.3–9.4 \times 10^3$
Nylon	493–508	0.65	$1.8–2.6 \times 10^3$
Polymethylmethyacrylate (PMMA)	493–533	0.25	$2.5–9 \times 10^4$
Polycarbonate	553–593	0.65–0.8	$1–8.5 \times 10^3$
Personal care products			
Nail polish	298	0.86	750
Mascara	298	0.24	200
Toothpaste	298	0.28	120
Sunscreen lotions	298	0.28	75
Ponds cold cream	298	0.45	25
Oil of Olay	298	0.22	25

Source: Modified after Johnson (1999)

Table 1. Typical values of power-law constants for a few systems

iii. The Cross viscosity equation

Another four parameter model which has gained wide acceptance is due to Cross (1965) which, in simple shear, is written as:

$$\frac{\mu - \mu_\infty}{\mu_0 - \mu_\infty} = \frac{1}{1 + k(\dot{\gamma}_{yx})^n} \tag{5}$$

Here n (<1) and k are two fitting parameters whereas μ_0 and μ_∞ are the limiting values of the apparent viscosity at low and high shear rates, respectively. This model reduces to the Newtonian fluid behaviour as $k \rightarrow 0$. Similarly, when $\mu \ll \mu_0$ and $\mu \gg \mu_\infty$, it reduces to the familiar power-law model, equation (3). Though initially Cross (1965) suggested that a constant value of $n = 2/3$ was adequate to approximate the viscosity data for many systems, it is now thought that treating the index, n, as an adjustable parameter offers considerable improvement over the use of the constant value of n (Barnes *et al.* , 1989).

iv. The Ellis fluid model

When the deviations from the power-law model are significant only at low shear rates, it is more appropriate to use the Ellis model. The three viscosity equations presented so far are examples of the form of equation (1). The three-constant Ellis model is an illustration of the inverse form. In simple shear, the apparent viscosity of an Ellis model fluid is given by:

$$\mu = \frac{\mu_0}{1 + (\tau_{yx} / \tau_{1/2})^{\alpha-1}} \tag{6}$$

In this equation, μ_0 is the zero shear viscosity and the remaining two constants α (> 1) and $\tau_{1/2}$ are adjustable parameters. While the index α is a measure of the degree of shear thinning behaviour (the greater the value of α , greater is the extent of shear-thinning), $\tau_{1/2}$ represents the value of shear stress at which the apparent viscosity has dropped to half its zero shear value. Equation (6) predicts Newtonian fluid behaviour in the limit of $\tau 1/2 \rightarrow \infty$. This form of equation has advantages in permitting easy calculation of velocity profiles from a known stress distribution, but renders the reverse operation tedious and cumbersome. It can easily be seen that in the intermediate range of shear stress (or shear rate), $(\tau_{yx} / \tau_{1/2})^{\alpha-1} \gg 1$, and equation (6) reduces to equation (3) with $n = (1/\alpha)$ and $m = (\mu_0 \tau_{1/2}^{\alpha-1})^{1/\alpha}$

2.2 Viscoplastic fluid behaviour

This type of fluid behaviour is characterized by the existence of a yield stress (τ_0) which must be exceeded before the fluid will deform or flow. Conversely, such a material will deform elastically (or flow *en masse* like a rigid body) when the externally applied stress is smaller than the yield stress. Once the magnitude of the external stress has exceeded the value of the yield stress, the flow curve may be linear or non-linear but will not pass through origin (Figure 1). Hence, in the absence of surface tension effects, such a material will not level out under gravity to form an absolutely flat free surface. One can, however, explain this kind of fluid behaviour by postulating that the substance at rest consists of three-dimensional structures of sufficient rigidity to resist any external stress less than τ_0. For stress levels greater than τ_0, however, the structure breaks down and the substance behaves like a viscous material. In some cases, the build-up and breakdown of structure has been found to be reversible, i.e., the substance may regain its initial value of the yield stress.

A fluid with a linear flow curve for $|\tau_{yx}| > |\tau_0|$ is called a Bingham plastic fluid and is characterized by a constant plastic viscosity (the slope of the shear stress versus shear rate curve) and a yield stress. On the other hand, a substance possessing a yield stress as well as a non-linear flow curve on linear coordinates (for $|\tau_{yx}| > |\tau_0|$), is called a yield pseudoplastic material. It is interesting to note that a viscoplastic material also displays an apparent viscosity which decreases with increasing shear rate. At very low shear rates, the apparent viscosity is effectively infinite at the instant immediately before the substance yields and begins to flow. It is thus possible to regard these materials as possessing a particular class of shear-thinning behaviour.

Strictly speaking, it is virtually impossible to ascertain whether any real material has a true yield stress or not, but nevertheless the concept of a yield stress has proved to be convenient in practice because some materials closely approximate to this type of flow behaviour, e.g. see Barnes and Walters (1985) , Astarita (1990) , Schurz (1990) and Evans (1992) . Many workers in this field view the yield stress in terms of the transition from a solid-like (high viscosity) to a liquid-like (low viscosity) state which occurs abruptly over an extremely narrow range of shear rates or shear stress (Uhlherr *et al* ., 2005). It is not uncommon for the two values of viscosity to differ from each other by several orders of magnitude. The answer to the question whether a fluid has a yield stress or not seems to be related to the choice of a time scale of observation. Common examples of viscoplastic fluid behaviour include particulate suspensions, emulsions, foodstuffs, blood and drilling mud, etc. (Barnes, 1999).

Over the years, many empirical expressions have been proposed as a result of straightforward curve-fitting exercises. A model based on sound theory is yet to emerge. Three commonly used models for viscoplastic fluids are: Bingham plastic model, Herschel-Bulkley model and Casson model.

i.   The Bingham plastic model

This is the simplest equation describing the flow behaviour of a fluid with a yield stress and, in steady one-dimensional shear, it is described by

$$\tau_{yx} = \tau_0 + \mu(\dot{\gamma}_{yx}) \text{ for } |\tau_{yx}| > |\tau_0|$$

$$\dot{\gamma}_{yx} = 0 \text{ for } |\tau_{yx}| < |\tau_0| \tag{7}$$

Often, the two model parameters τ_0 and μ are treated as curve-fitting constants irrespective of whether or not the fluid possesses a true yield stress.

ii.   The Herschel-Bulkley fluid model

A simple generalization of the Bingham plastic model to embrace the non-linear flow curve (for $\tau_{yx} > \tau_0$) is the three constant Herschel–Bulkley fluid model. In one dimensional steady shearing motion, the model is written as:

$$\tau_{yx} = \tau_0 + K(\dot{\gamma}_{yx})^n \text{ for } \left|\tau_{yx}\right| > \left|\tau_0\right|$$

$$\dot{\gamma}_{yx} = 0 \text{ for } \left|\tau_{yx}\right| < \left|\tau_0\right| \tag{8}$$

It is again noted that, the dimensions of K depend upon the value of n. With the use of the third parameter, this model provides a somewhat better fit to some experimental data.

iii.   The Casson fluid model

Many foodstuffs and biological materials, especially blood, are well described by this two constant model as:

$$\left|\tau_{yx}\right|^{1/2} = \left|\tau_0\right|^{1/2} + (\mu / \dot{\gamma}_{yx})^{1/2} \text{ for } \left|\tau_{yx}\right| > \left|\tau_0\right|$$

$$\dot{\gamma}_{yx} = 0 \text{ for } \left|\tau_{yx}\right| < \left|\tau_0\right| \tag{9}$$

This model has often been used for describing the steady shear stress–shear rate behaviour of blood, yoghurt, tomato purée, molten chocolate, etc. The flow behaviour of some particulate suspensions also closely approximates to this type of behaviour. The comparative performance of these three as well as several other models for viscoplastic behaviour has been thoroughly evaluated in an extensive review paper by Bird *et al* . (1983) and a through discussion on the existence, measurement and implications of yield stress has been provided by Barnes (1999).

2.3 Shear-thickening or dilatant fluid behaviour

Dilatant fluids are similar to pseudoplastic systems in that they show no yield stress but their apparent viscosity increases with increasing shear rate; thus these fluids are also called shear-thickening. This type of fluid behaviour was originally observed in concentrated suspensions and a possible explanation for their dilatant behaviour is as follows: At rest, the voidage is minimum and the liquid present is sufficient to fill the void space. At low shear rates, the liquid lubricates the motion of each particle past others and the resulting stresses are consequently small. At high shear rates, on the other hand, the material expands or dilates slightly (as also observed in the transport of sand dunes) so that there is no longer sufficient liquid to fill the increased void space and prevent direct solid–solid contacts which result in increased friction and higher shear stresses. This mechanism causes the apparent viscosity to rise rapidly with increasing rate of shear. The term dilatant has also been used for all other fluids which exhibit increasing apparent viscosity with increasing rate of shear. Many of these, such as starch pastes, are not true suspensions and show no dilation on shearing. The above explanation therefore is not applicable but nevertheless such materials are still commonly referred to as dilatant fluids.

Of the time-independent fluids, this sub-class has received very little attention; consequently very few reliable data are available. Until recently, dilatant fluid behaviour was considered to be much less widespread in the chemical and processing industries. However, with the recent growing interest in the handling and processing of systems with high solids loadings, it is no longer so, as is evidenced by the number of recent review articles on this subject (Barnes *et al.*, 1987; Barnes, 1989; Goddard and Bashir, 1990). Typical examples of materials exhibiting dilatant behaviour include concentrated suspensions of china clay, titanium dioxide (Metzner and Whitlock, 1958) and of corn fl our in water (Griskey *et al.*, 1985). The limited information reported so far suggests that the apparent viscosity–shear rate data often result in linear plots on double logarithmic coordinates over a limited shear rate range and the flow behaviour may be represented by the power-law model, with the flow behaviour index, n, greater than unity, i.e.,

$$\mu = K(\dot{\gamma}_{xy})^{n-1} \tag{10}$$

One can readily see that for $n > 1$, equation (10) predicts increasing viscosity with increasing shear rate. The dilatant behaviour may be observed in moderately concentrated suspensions at high shear rates, and yet, the same suspension may exhibit pseudoplastic behaviour at lower shear rates.

This section is concluded by Table 2 providing a list of materials displaying a spectrum of non-Newtonian flow characteristics in diverse applications to reinforce idea yet again of the ubiquitous nature of such flow behaviour.

Practical fluid	Characteristics	Consequence of non-Newtonian behaviour
Toothpaste	Bingham Plastic	Stays on brush and behaves more liquid-like while brushing
Drilling muds	Bingham Plastic	Good lubrication properties and ability to convey debris
Non-drip paints	Thixotropic	Thick in the tin, thin on the brush
Wallpaper paste	Pseudoplastic and Viscoelastic	Good spreadability and adhesive properties
Egg white	Visco-elastic	Easy air dispersion (whipping)
Molten polymers	Visco-elastic	Thread-forming properties
' Bouncing Putty '	Visco-elastic	Will fl ow if stretched slowly, but will bounce (or shatter) if hit sharply
Wet cement aggregates	Dilatant and thixotropic	Permit tamping operations in which small impulses produce almost complete settlement
Printing inks	Pseudoplastic	Spread easily in high speed machines yet do not run excessively at low speeds
Waxy crude oils	Viscoplastic and Thixotropic	Flows readily in a pipe, but difficult to restart the flow

Table 2. Non-Newtonian characteristics of some common materials

3. Rabinowitsch-Mooney equation

Consider a one-directional flow of fluid through a circular tube with radius R, Figure 2. The volumetric flow rate through an annular element of area perpendicular to the flow and of width δr is given by

$$\delta Q = 2\pi r \delta r . v_x \tag{11}$$

and, consequently, the flow rate through the whole tube is

$$Q = 2\pi \int_0^R r v_x dr \tag{12}$$

Integrating by parts gives

$$Q = 2\pi \left\{ \left[\frac{r^2 v_x}{2} \right]_0^{r_i} + \int_0^{r_i} \frac{r^2}{2} \left(-\frac{dv_x}{dr} \right) dr \right\} \tag{13}$$

Provided there is no slip at the tube wall, the first term in equation (13) vanishes. Equation (13) then can be written as

$$Q = \pi \int_0^R r^2 (-\dot{\gamma}) dr \tag{14}$$

If the fluid is time-independent and homogeneous, the shear stress is a function of shear rate only. The inverse is that the shear rate $\dot{\gamma}$, is a function of shear stress τ_{rx} only and the variation of τ_{rx} with r is known from the following well-known equation:

$$\frac{\tau_{rx}}{\tau_w} = \frac{r}{R} \tag{15}$$

where τ_w is the wall shear stress.

Changing variables in equation (14), using equation (15), and dropping the subscripts rx, equation (14) can be written as

$$Q = \pi \int_0^{\tau_w} \frac{\tau^2 R^2}{\tau_w^2} (-\dot{\gamma}) \frac{\tau_i}{\tau_w} d\tau = \frac{\tau R^3}{\tau_w^3} \int_0^{\tau_w} \tau^2 (-\dot{\gamma}) d\tau \tag{16}$$

where $\dot{\gamma}$ is interpreted as a function of τ instead of r.

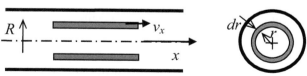

Fig. 2. Geometric presentation of MR fluid in o circular tube

Writing equation (16) in terms of the flow characteristic gives

$$\frac{8u}{D} = \frac{4Q}{\pi R^3} = \frac{4}{\tau_w^3}\int_0^{\tau_w} \tau^2(-\dot{\gamma})d\tau \tag{17}$$

where u is the average velocity of the fluid flow and D is the diameter of the tube. For flow in a pipe or tube the shear rate is negative so the integral in equation (17) is positive. For a given relationship between τ and $\dot{\gamma}$, the value of the integral depends only on the value of τ_w. Thus, for a non-Newtonian fluid, as well as for a Newtonian fluid, the flow characteristic $8u/D$ is a unique function of the wall shear stress τ_w.

The shear rate $\dot{\gamma}$ can be extracted from equation (17) by differentiating with respect to τ. Moreover, if a definite integral is differentiated w.r.t. the upper limit (τ_w), the result is the integrand evaluated at the upper limit. It is convenient first to multiply equation (17) by τ_w^3 throughout, then differentiating w.r.t. τ_w gives

$$3\tau_w^2\frac{8u}{D} + \tau_w^3\frac{d(8u/D)}{d\tau_w} = 4\tau_w^2(-\dot{\gamma})_w \tag{18}$$

Rearranging equation (18) gives the wall shear rate $\dot{\gamma}_w$ as

$$-\dot{\gamma}_w = \frac{8u}{D}\left[\frac{3}{4} + \frac{1}{4}\frac{\tau_w}{(8u/D)}\frac{d(8u/D)}{d\tau_w}\right] \tag{19}$$

Making use of the relationship $dx/x = dlnx$, equation (19) can be written as

$$-\dot{\gamma}_w = \frac{8u}{D}\left[\frac{3}{4} + \frac{1}{4}\frac{d\ln(8u/D)}{d\ln\tau_w}\right] \tag{20}$$

As the wall shear rate $\dot{\gamma}_{wN}$ for a Newtonian fluid in laminar flow is equal to $(-8u/D)$, equation (20) can be expressed as

$$\dot{\gamma}_w = \dot{\gamma}_{wN}\left[\frac{3}{4} + \frac{1}{4}\frac{d\ln(8u/D)}{d\ln\tau_w}\right] \tag{21}$$

Equations (20) and (21) are forms of the Rabinowitsch-Mooney equation. It shows that the wall shear rate for a non-Newtonian fluid can be calculated from the value for a Newtonian fluid having the same flow rate in the same pipe, the correction factor being the quantity in the square brackets. The derivative can be estimated by plotting $ln(8u/D)$ against $\ln\tau_w$ and measuring the gradient. Alternatively the gradient may be calculated from the (finite) differences between values of $ln(8du/D)$ and $\ln\tau_w$. Thus the flow curve τ_w against $\dot{\gamma}_w$ can be determined. The measurements required and the calculation procedure are as follows.

1. Measure Q at various values of $\Delta P_f/L$, preferably eliminating end effects.
2. Calculate τ_ω from the pressure drop measurements and the corresponding values of the flow characteristic $(8du/D = 4Q/\pi R^3)$ from the flow rate measurements.

3.   Plot $\ln(8du/D)$ against $\ln\tau_w$ and measure the gradient at various points on the curve. Alternatively, calculate the gradient from the differences between the successive values of these quantities.
4.   Calculate the true wall shear rate from equation (20) with the derivative determined in step 3. In general, the plot of $\ln(8du/D)$ against $\ln\tau_w$ will not be a straight line and the gradient must be evaluated at the appropriate points on the curve.

Example 1

The flow rate-pressure drop measurements shown in Table 3 were made in a horizontal tube having an internal diameter D = 6 mm, the pressure drop being measured between two tapings 2.0m apart. The density of the fluid, ρ, was 870 kg/m³. Determine the wall shear stress-flow characteristic curve and the shear stress-true shear rate curve for this material.

Pressure drop (bar)	Mass flow rate x 10³ (kg/s)
0.384	0.0864
0.519	0.463
0.716	1.37
0.965	2.76
1.16	4.13
1.29	5.20
1.46	6.78
1.60	8.15

Table 3.

The results are shown in Table 4

τ_w (Pa)	$8du/D$	gradient n'	$(3n'+1)/4n'$	$-\dot\gamma_w\,(s^{-1})$
28.8	4.68	0.157	2.34	11.0
38.9	25.1	0.232	1.83	45.9
53.7	74.3	0.375	1.42	106
72.4	150	0.439	1.32	197
87.0	224	0.475	1.28	286
96.8	282	0.475	1.28	360
110	367	0.475	1.28	469
120	442	0.475	1.28	564

Table 4

4. Calculation of flow rate-pressure drop relationship for laminar flow using $\tau-\dot\gamma$ data

Flow rate-pressure drop calculations for laminar non-Newtonian flow in pipes may be made in various ways depending on the type of flow information available. When the flow data are in the form of flow rate and pressure gradient measured in a tubular viscometer or in a pilot scale pipeline, direct scale-up can be done as described in Section 5. When the data are in the form of shear stress-shear rate values (tabular or graphical), the flow rate can be calculated directly using equation (17), where D is the diameter of the pipe to be used and τ_w is the wall shear stress corresponding to the specified pressure gradient. Whether obtained with a rotational instrument or with a tubular viscometer, the data provide the relationship between τ and $\dot{\gamma}$. Numerical evaluation of the integral in equation (17) can be done using selected pairs of values of τ and $\dot{\gamma}$ ranging from 0 to τ_w.

If the $\tau - \dot{\gamma}$, relationship can be accurately represented by a simple algebraic expression, such as the power law, over the required range then this may be used to substitute for $\dot{\gamma}$, in equation (17), allowing the integral to be evaluated analytically. Both these methods are illustrated in the following example.

Example 2

Using the viscometric data given in Table 5 calculate the average velocity for the material flowing through a pipe of diameter 37mm when the pressure gradient is 1.1kPa/m.

$\dot{\gamma}(s^{-1})$	τ (Pa)	μ_a (Pa s)
0.00911	0.0417	4.58
0.0911	0.175	1.95
0.911	0.708	0.777
9.111	2.82	0.310
91.11	11.22	0.123
102.3	12.03	0.118

Table 5.

Calculations

The wall shear stress is given by

$$\tau_w = \frac{D\Delta P}{4L} = \frac{(37 \times 10^{-3}\text{m})(1100 \text{ Pa/m})}{4} = 10.18 \text{ Pa}$$

the flow characteristic

$$\frac{8u}{D} = \frac{4}{\tau_w^3}\int_0^{\tau_w} \tau^2(-\dot{\gamma})d\tau$$

It is necessary to evaluate the integral from $\tau = 0$ to $\tau = 10.18Pa$. This can be done by calculating $\tau^2\dot{\gamma}$ for each of the values given in the table and plotting $\tau^2\dot{\gamma}$ against τ. The area under the curve between $\tau = 0$ and $\tau = 10.18Pa$ can then be measured. An alternative, which will be used here, is to use a numerical method such as Simpson's rule. This requires values at equal intervals of τ. Dividing the range of integration into six strips and interpolating the data allows Table 6 to be constructed.

τ (Pa)	$\dot{\gamma}$	$\tau^2\dot{\gamma}$
0.00	0.0	0.00 (Centerline)
1.70	3.91	11.24
3.39	12.41	142.8
5.09	24.38	631.0
6.78	39.39	1812
8.48	57.14	4108
10.18	77.43	8016 (pipe wall)

Table 6.

By Simpson's rule

$$\int_0^{10.18} \tau^2\dot{\gamma}\,d\tau \approx \frac{10.18/6}{3}[0+8016+4(11.24+613+4108)+2(142.8+1812)]= \ 17490 \ Pa^3s^{-1}$$

From equation (17)

$$u = \frac{(37 \times 10^{-3}\text{m})(17490 \ Pa^3s^{-1})}{2(10.18 \ Pa)^3} = 0.307 \ \text{m/s}$$

The above is the general method but in this case the viscometric data can be well represented by $\tau = 0.749\dot{\gamma}^{0.60}$ Pa, thus $\dot{\gamma} = 1.62\tau^{1.667}$ s⁻¹. This allows the integral in equation (17) to be evaluated analytically.

$$\int_0^{\tau_\omega} \tau^2\dot{\gamma}\,d\tau = 1.62 \int_0^{10.18} \tau^{3.667}\,d\tau = 17510 \ Pa^3s^{-1}$$

This agrees with the value found by numerical integration and would give the same value for u.

Note that the values of the apparent viscosity μ_0 were not used; they were provided to show that the fluid is strongly shear thinning. If the data were available as values of μ_0 at corresponding values of $\dot{\gamma}$, then τ should be calculated as their product. The table of values of $\tau^2\dot{\gamma}$ (Table 6) illustrates the fact that flow in the centre makes a small contribution to the total flow: flow in the outer parts of the pipe is most significant.

As mentioned previously, the minus sign in equation (17) reflects the fact that the shear rate is negative for flow in a pipe. In the above calculations, the absolute values of $\dot{\gamma}$, and τ have been used and the minus sign has therefore been dropped.

5. Wall shear stress-flow characteristic curves and scale-up for laminar flow

When data are available in the form of the flow rate-pressure gradient relationship obtained in a small diameter tube, direct scale-up for flow in larger pipes can be done. It is not necessary to determine the τ - $\dot{\gamma}$ curve with the true value of $\dot{\gamma}$ calculated from the

Rabinowitsch-Mooney equation (Equation (20)). Equation (17) shows that the flow characteristic is a unique function of the wall shear stress for a particular fluid:

$$\frac{8u}{D} = \frac{4}{\tau_w^3} \int_0^{\tau_w} \tau^2 (-\dot{\gamma}) d\tau$$

In the case of a Newtonian fluid, substituting $\dot{\gamma} = -\tau / \mu$ into the above equation and evaluating the integral gives

$$\frac{8u}{D} = \frac{\tau_w}{\mu} \qquad (22)$$

Recall that the wall shear rate for a Newtonian fluid in laminar flow in a tube is equal to $-8u / D$. In the case of a non-Newtonian fluid in laminar flow, the flow characteristic is no longer equal to the magnitude of the wall shear rate. However, the flow characteristic is still related uniquely to τ_w because the value of the integral, and hence the right hand side of equation (17), is determined by the value of τ_w.

If the fluid flows in two pipes having internal diameters D_1 and D_2 with the same value of the wall shear stress in both pipes, then from equation (17) the values of the flow characteristic are equal in both pipes:

$$\frac{8u_1}{D_1} = \frac{8u_2}{D_2} \qquad (23)$$

So the average velocities are related by

$$\frac{u_1}{u_2} = \frac{D_1}{D_2} \qquad (24)$$

By substituting for u or by writing the flow characteristic as $-4Q / \pi R^3$, the volumetric flow rates are related by

$$\frac{Q_1}{Q_2} = \left(\frac{D_1}{D_2} \right)^3 \qquad (25)$$

It is important to appreciate that the same value of τ_w requires different values of the pressure gradient in the two pipes. It is convenient to represent the flow behaviour as a graph of τ_w plotted against $8u / D$, as shown in Figure 3. In accordance with the above discussion, all data fit a single line for laminar flow. The graph is steeper for turbulent flow and different lines are found for different pipe diameters. It is noteworthy that the same would be found for Newtonian flow if the data were plotted in this way and the laminar flow line would be a straight line of gradient μ passing through the origin. The plot in Figure 3 is not a true flow curve because the flow characteristic is equal to the magnitude of the wall shear rate only in the case of Newtonian laminar flow.

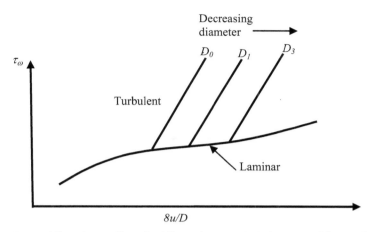

Fig. 3. Shear stress at the pipe wall against flow characteristic for a non-Newtonian fluid flowing in a pipe

Given a wall shear stress-flow characteristic curve such as that in Figure 3, the flow rate-pressure drop relationship can be found for any diameter of pipe provided the flow remains laminar and is within the range of the graph. For example, if it is required to calculate the pressure drop for flow in a pipe of given diameter at a specified volumetric flow rate, the value of the flow characteristic $(8u/D = 4Q/\pi R^3)$ is calculated and the corresponding value of the wall shear stress τ_w read from the graph. The pressure gradient, and hence the pressure drop for a given pipe length, can then be calculated.

It is found useful to define two quantifies K' and n' in order to describe the τ_w -flow characteristic curve. If the laminar flow data are plotted on logarithmic axes as in Figure 4, then the gradient of the curve defines the value of n' :

$$n' = \frac{d\ln \tau_w}{d\ln(8u/D)} \tag{26}$$

The equation of the tangent can be written as

$$\tau_w = K'\left(\frac{8u}{D}\right)^{n'} \tag{27}$$

In general, both K' and n' have different values at different points along the curve. The values should be found at the point corresponding to the required value of τ_w. In some cases, the curve in Figure 4 will be virtually straight over the range required and a single value may be used for each of K' and n'. Although equation (27) is similar to the equation of a power law fluid, the two must not be confused.

The reason for defining n' in this way can be seen from equation (21) where the inverse of the derivative occurs in the correction factor. Equation (20) can be written in terms of n' as

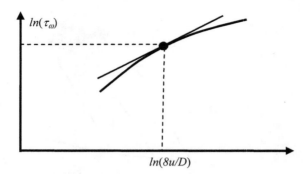

Fig. 4. Logarithmic plot of wall shear stress against flow characteristic: the gradient at a point defines n'

$$\dot{\gamma}_\omega = \dot{\gamma}_{\omega N}\left(\frac{3n'+1}{4n'}\right) \tag{28}$$

Equation (28) is helpful in showing how the value of the correction factor in the Rabinowitsch-Mooney equation corresponds to different types of flow behaviour. For a Newtonian fluid, $n' = 1$ and therefore the correction factor has the value unity. Shear thinning behaviour corresponds to $n' < 1$ and consequently the correction factor has values greater than unity, showing that the wall shear rate $\dot{\gamma}_w$ is of greater magnitude than the value for Newtonian flow. Similarly, for shear thickening behaviour, $\dot{\gamma}_w$ is of a smaller magnitude than the Newtonian value $\dot{\gamma}_{wN}$. The value correction factor varies from 2.0 for $n' = 0.2$ to 0.94 for $n' = 1.3$.

6. Generalized Reynolds number for flow in pipes

It is recalled that for Newtonian flow in a pipe, the Reynolds number is defined by

$$Re = \frac{\rho u D}{\mu} \tag{29}$$

In the case of non-Newtonian flow, it is necessary to use an appropriate apparent viscosity. Although the apparent viscosity μ_a is defined in the same way as for a Newtonian fluid, it no longer has the same fundamental significance and other, equally valid, definitions of apparent viscosities may be made. In flow in a pipe, where the shear stress varies with radial location, the value of μ_a varies. It is shown that the conditions near the pipe wall that are most important. The value of μ_a evaluated at the wall is given by

$$\mu_a = -\frac{\text{shear stress at wall}}{\text{shear rate at wall}} = \frac{\tau_\omega}{(-dv_x / dr)_\omega} \tag{30}$$

Another definition is based, not on the true shear rate at the wall, but on the flow characteristic. This quantity, which may be called the apparent viscosity for pipe flow, is given by

$$\mu_{ap} = \frac{\text{shear stress at wall}}{\text{flow characteristic}} = \frac{\tau_\omega}{(8u / d_i)} \tag{31}$$

For laminar flow, μ_{ap} has the property that it is the viscosity of a Newtonian fluid having the same flow characteristic as the non-Newtonian fluid when subjected to the same value of wall shear stress. In particular, this corresponds to the same volumetric flow rate for the same pressure gradient in the same pipe. This suggests that μ_{ap} might be a useful quantity for correlating flow rate-pressure gradient data for non-Newtonian flow in pipes. This is found to be the case and it is on μ_{ap} that a generalized Reynolds number Re' is based

$$Re' = \frac{\rho u d_i}{\mu_{ap}} \tag{32}$$

Representing the fluid's laminar flow behaviour in terms of K' and n'

$$\tau_\omega = K'\left(\frac{8u}{d_i}\right)^{n'} \tag{33}$$

The pipe flow apparent viscosity, defined by equation 31, is given by

$$\mu_{ap} = \frac{\tau_\omega}{8u / d_i} = K'\left(\frac{8u}{d_i}\right)^{n'-1} \tag{34}$$

Using μ_{ap} in Equation (34), the generalized Reynolds number takes the form

$$Re' = \frac{\rho u^{2-n'} d_i^{n'}}{8^{n'-1} K'} \tag{35}$$

Use of this generalized Reynolds number was suggested by Metzner and Reed (1955). For Newtonian behaviour, $K' = \mu$ and $n' = 1$ so that the generalized Reynolds number reduces to the normal Reynolds number.

7. Turbulent flow of Inelastic non-Newtonian fluids in pipes and circular ducts

Turbulent flow of Newtonian fluids is described in terms of the Fanning friction factor, which is correlated against the Reynolds number with the relative roughness of the pipe wall as a parameter. The same approach is adopted for non-Newtonian flow but the generalized Reynolds number is used. The Fanning friction factor is defined by

$$f = \frac{\tau_\omega}{\frac{1}{2}\rho u^2} \tag{36}$$

It is straightforward to show that the Fanning friction factor for laminar non-Newtonian flow becomes

$$f = 16 / Re' \tag{37}$$

This is of the same form as equation for Newtonian flow and is one reason or using this form of generalized Reynolds number. Equation (37) provides another way of calculating the pressure gradient for a given flow rate for laminar non-Newtonian flow.

7.1 Laminar-turbulent transition

A stability analysis made by Ryan and Johnson (1959) suggests that the transition from laminar to turbulent flow for inelastic non-Newtonian fluids occurs at a critical value of the generalized Reynolds number that depends on the value of n'. The results of this analysis are shown in Figure 5. This relationship has been tested for shear thinning and for Bingham plastic fluids and has been found to be accurate. Over the range of shear thinning behaviour encountered in practice, $0.2 \leq n' \leq 1$, the critical value of Re' is in the range $2100 \leq Re' \leq 2400$.

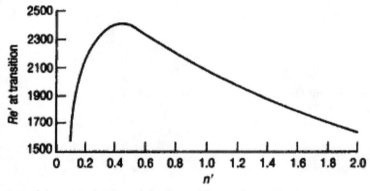

Fig. 5. Variation of the critical value of the Reynolds number with n'

7.2 Friction factors for turbulent flow in smooth pipes

Experimental results for the Fanning friction factor for turbulent flow of shear thinning fluids in smooth pipes have been correlated by Dodge and Metzner (1959) as a generalized form of the yon Kármán equation:

$$\frac{1}{f^{1/2}} = \frac{4.0}{(n')^{0.75}} \log[f^{1-n'/2}Re'] - \frac{0.40}{(n')^{1.2}} \tag{38}$$

This correlation is shown in Figure 6. The broken lines represent extrapolation of equation (38) for values of n' and Re' beyond those of the measurements made by Dodge and Metzner. More recent studies tend to confirm the findings of Dodge and Metzner but do not significantly extend the range of applicability. Having determined the value of the friction factor f for a specified flow rate and hence Re', the pressure gradient can be calculated in the normal way.

Example 3

A general time-independent non-Newtonian liquid of density 961 kg/m³ flows steadily with an average velocity of 2.0 m/s through a tube 3.048 m long with an inside diameter of 0.0762 m. For these conditions, the pipe flow consistency coefficient K' has a value of 1.48 Pa s$^{0.3}$

and n' a value of 0.3. Calculate the values of the apparent viscosity for pipe flow μ_{ap}, the generalized Reynolds number Re' and the pressure drop across the tube, neglecting end effects.

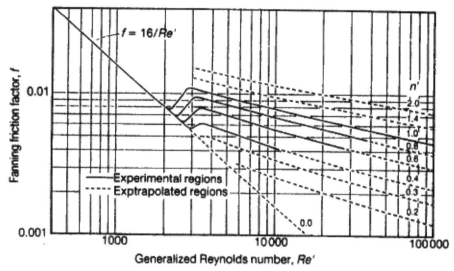

Source: D. W. Dodge and A. B. Metzner, *AIChE Journal* **5** (1959) 189-204

Fig. 6. Friction factor chart for purely viscous non-Newtonian fluids.

Calculations

The flow characteristic is given by

$$\frac{8u}{D} = \frac{8(2.0m/s)}{0.0762m} = 210s^{-1}$$

and

$$\left(\frac{8u}{D}\right)^{n'-1} = 210^{(0.3-1.0)} = 0.0237s^{0.7}$$

Hence

$$\mu_{ap} = K'\left(\frac{8u}{D}\right)^{n'-1} = (1.48 \text{ Pa s}^{0.3})(0.0237 \text{ s}^{0.7}) = 0.0351 \text{ Pa s}$$

and

$$Re' = \frac{\rho u D}{\mu_{ap}} = \frac{(0.0762 \text{ m})(2.0 \text{ m})(961 \text{ kg/m}^3)}{(0.0351 \text{ Pa s})} = 4178$$

From Figure 6, the Fanning friction factor f has a value 0.0047. Therefore the pressure drop is given by

$$\Delta P_f = 4f\left(\frac{L}{d_i}\right)\frac{\rho u^2}{2} = \frac{2(0.0047)(3.048 \text{ m})(961 \text{ kg/m}^3)(2.0 \text{ m/s})^2}{(0.0762 \text{ m})} = 1445 \text{ Pa}$$

8. Laminar flow of inelastic fluids in non-circular ducts

Analytical solutions for the laminar flow of time-independent fluids in non-axisymmetric conduits are not possible. Numerous workers have obtained approximate and/or complete numerical solutions for specific flow geometries including square, rectangular and triangular pipes (Schechter, 1961 ; Wheeler and Wissler, 1965 ; Miller, 1972 ; Mitsuishi and Aoyagi, 1969, 1973). On the other hand, semi-empirical attempts have also been made to develop methods for predicting pressure drop for time-independent fluids in ducts of non-circular cross-section. Perhaps the most systematic and successful friction factor analysis is that provided by Kozicki *et al* . (1966, 1967) . It is useful to recall here that the equation (19) is a generalized equation for the laminar flow of time-independent fluids in a tube and it can be slightly rearranged as:

$$-\dot{\gamma}_w = f(\tau_w) = \frac{1}{4}\tau_w \frac{d(8u/D)}{d\tau_w} + \frac{3}{4}(\frac{8u}{D}) \tag{39}$$

Similarly, one can parallel this approach for the fully developed laminar flow of time independent fluids in a thin slit (Figure 7) to derive the following relationship:

$$-\dot{\gamma}_w = \left(-\frac{dV_z}{dr}\right)_w = f(\tau_w) = \tau_w \frac{d(u/h)}{d\tau_w} + 2(\frac{u}{h}) \tag{40}$$

In order to develop a unified treatment for ducts of various cross-sections, it is convenient to introduce the usual hydraulic diameter D_h (defined as four times the area for flow/wetted perimeter) into equations (39) and 40).

For a circular pipe, $D_h = D$ and hence equation (38) becomes:

$$-\dot{\gamma}_\omega = f(\tau_\omega) = \frac{1}{4}\tau_\omega \frac{d(8u/D_h)}{d\tau_\omega} + \frac{3}{4}(\frac{8u}{D_h}) \tag{41}$$

For the slit shown in Figure 7, the hydraulic diameter $D_h = 4h$, and thus equation (39) is rewritten as:

$$-\dot{\gamma}_w = \frac{1}{2}\tau_w \frac{d(8u/D_h)}{d\tau_w} + (\frac{8u}{D_h}) \tag{42}$$

By noting the similarity between the form of the Rabinowitsch–Mooney equations for the flow of time-independent fluids in circular pipes (equation (41)) and that in between two plates (equation (42)), they suggested that it could be extended to the ducts having a constant cross-section of arbitrary shape as follows:

$$-\dot\gamma_w = f(\overline\tau_w) = a\overline\tau_w \frac{d(8u/D_h)}{d\overline\tau_w} + b(\frac{8u}{D_h})$$ (43)

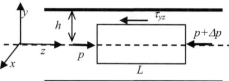

Fig. 7. Laminar flow between parallel plates

where a and b are two geometric parameters characterizing the cross-section of the duct ($a=1/4$ and $b= 3/4$ for a circular tube, and $a =1/2$ and $b =1$ for the slit) and $\overline\tau_w$ is the mean value of shear stress at the wall, and is related to the pressure gradient as:

$$\overline\tau_w = \frac{D_h}{4}\left(\frac{-\Delta P}{L}\right)$$ (44)

For constant values of a and b , equation (42) is an ordinary differential equation of the form $(d\,y/d\,x) + p(x)\,y = q(x)$ which can be integrated to obtain the solution as:

$$y = e^{-\int p(x)dx}\int e^{\int p(x)dx}q(x)dx + C_0$$ (45)

Now identifying $y= (8V/D_h)$ and $x =\overline\tau_w$, $p(x) =(b\,/a\overline\tau_w)$ and $q(x)= (f(\overline\tau_w)/a\overline\tau_w)$, the solution to equation (43) is given as:

$$\frac{8u}{D_h} = \frac{1}{a}(\overline\tau_w)^{-b/a}\int_0^{\overline\tau_w}\xi^{(b/a)-1}f(\tau)d\xi$$ (46)

where ξ is a dummy variable of integration. The constant C_0 has been evaluated by using the condition that when $V=0$, $\overline\tau_w = 0$ and therefore, $C_0 =0$.

For the flow of a power-law fluid, $f(\tau) = (\tau/K)^{1/n}$ and integration of equation (46) yields:

$$\overline\tau_w = K\left\{\frac{8u}{D_n}(b+\frac{a}{n})\right\}^n$$ (47)

which can be rewritten in terms of the friction factor, $f= 2\overline\tau_w /\rho u^2$ as:

$$f = \frac{16}{Re_g}$$ (48)

where the generalized Reynolds number,

$$Re_g = \frac{\rho u^{2-n}D_h^n}{8^{n-1}K(b+\frac{a}{n})^n}$$ (49)

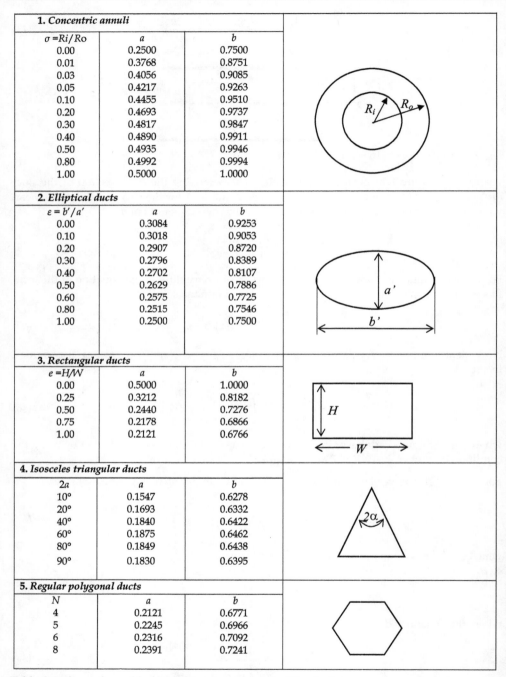

1. Concentric annuli

$\sigma = Ri/Ro$	a	b
0.00	0.2500	0.7500
0.01	0.3768	0.8751
0.03	0.4056	0.9085
0.05	0.4217	0.9263
0.10	0.4455	0.9510
0.20	0.4693	0.9737
0.30	0.4817	0.9847
0.40	0.4890	0.9911
0.50	0.4935	0.9946
0.80	0.4992	0.9994
1.00	0.5000	1.0000

2. Elliptical ducts

$\varepsilon = b'/a'$	a	b
0.00	0.3084	0.9253
0.10	0.3018	0.9053
0.20	0.2907	0.8720
0.30	0.2796	0.8389
0.40	0.2702	0.8107
0.50	0.2629	0.7886
0.60	0.2575	0.7725
0.80	0.2515	0.7546
1.00	0.2500	0.7500

3. Rectangular ducts

$e = H/W$	a	b
0.00	0.5000	1.0000
0.25	0.3212	0.8182
0.50	0.2440	0.7276
0.75	0.2178	0.6866
1.00	0.2121	0.6766

4. Isosceles triangular ducts

$2a$	a	b
10°	0.1547	0.6278
20°	0.1693	0.6332
40°	0.1840	0.6422
60°	0.1875	0.6462
80°	0.1849	0.6438
90°	0.1830	0.6395

5. Regular polygonal ducts

N	a	b
4	0.2121	0.6771
5	0.2245	0.6966
6	0.2316	0.7092
8	0.2391	0.7241

Table 6. Values of a and b depending on geometry of the ducts

The main virtue of this approach lies in its simplicity and the fact that the geometric parameters a and b can be deduced from the behaviour of Newtonian fluids in the same flow geometry. Table 7 lists values of a and b for a range of flow geometries commonly encountered in process applications.

Kozicki et al.(1966) argued that the friction factor of the turbulent flow in non-circular ducts can be calculated from the following equation

$$\frac{1}{\sqrt{f}} = \frac{4}{n^{0.75}}\log_{10}(\text{Re}_g \, f^{(2-n)/2}) - \frac{0.4}{n^{1.2}} + 4n^{0.25}\log[\frac{a(a+bn)}{3n+1}] \tag{50}$$

Note that since for a circular tube, $a = 1/4$ and $b = 3/4$, equation (50) is consistent with that for circular pipes. The limited data available on turbulent flow in triangular (Irvine Jr, 1988), rectangular (Kostic and Hartnett, 1984) and square ducts (Escudier and Smith, 2001) conforms to equation (48). In the absence of any definite information, Kozicki and Tiu (1988) suggested that the Dodge–Metzner criterion, $\text{Re}_g \leq 2100$, can be used for predicting the limit of laminar flow in non-circular ducts.

Some further attempts have been made to simplify and/or improve upon the two geometric parameter method of Kozicki et al . (1966, 1967). Delplace and Leuliet (1995) revisited the definition of the generalized Reynolds number (equation (49)) and argued that while the use of a and b accounts for the non-circular cross-sections of the ducts, but the factor 8^{n-1} appearing in the denominator is strictly applicable for the flow in circular ducts only. Their reasoning hinges on the fact that for the laminar flow of a Newtonian fluid, the product $(f.Re)$ is a function of the conduit shape only. Thus, they wrote

$$\beta = \frac{48}{(f.Re)} \tag{51}$$

where both the (Fanning) friction factor and the Reynolds number are based on the use of the hydraulic diameter, D_h and the mean velocity of the flow, u. Furthermore, they were able to link the geometric parameters a and b with the new parameter β as follows:

$$a = \frac{1}{1+\beta}; \quad b = \frac{\beta}{1+\beta} \tag{52}$$

and finally, the factor of 8^{n-1} in the denominator in equation (49) is replaced $(24/\beta)^{n-1}$. With these modifications, one can use the relationship $f = (16/Re_g)$ to estimate the pressure gradient for the laminar flow of a power-law fluid in a non-circular duct for which the value of β is known either from experiments or from numerical results. Therefore, this approach necessitates the knowledge of only one parameter (β) as opposed to the two geometric parameters, namely, a and b in the method of Kozicki et al . (1966) and Kozicki and Tiu (1967), albeit a similar suggestion was also made by Miller (1972) and Liu (1983). Finally, for the limiting case of a circular pipe, evidently $\beta=3$ thereby leading to $a=(1/4)$ and $b=(3/4)$ and the two definitions of the Reynolds number coincide, as expected. The values of β for a few standard duct shapes are summarized in Table 5. While in laminar flow, these two methods give almost identical predictions, the applicability of the modified method of Delplace and Leuliet (1995) has not been checked in the transitional and turbulent flow regions. Scant analytical and experimental results suggest that visco-elasticity in a fluid may induce secondary motion in non-circular conduits, even under laminar conditions. However, measurements reported to date indicate that the friction factor–Reynolds number behaviour is little influenced by such secondary flows (Hartnett and Kostic, 1989).

Example 4

A power-law fluid (K = 0.3 Pa.s^n and n = 0.72) of density 1000 kg/m³ is flowing in a series of ducts of the same flow area but different cross-sections as listed below:

i.   concentric annulus with R= 37mm and $\sigma=(R/R_i)$= 0.40
ii.   circular pipe of radius R
iii. rectangular, (H / W) =0.5
iv. elliptical, b'/a'= 0.5

Estimate the pressure gradient required to maintain an average velocity of 1.25m/s in each of these channels. Use the geometric parameter method. Also, calculate the value of the generalized Reynolds number as a guide to the nature of the flow.

Solution

i.   For a concentric annulus, $\sigma = 0.4$

-     From Table 6 we have: a =0.489; b = 0.991
-     The hydraulic diameter, D_h = $2R(1-\sigma)$=0.044m.
-     Reynolds number, $\mathrm{Re}_g = \dfrac{\rho u^{2-n}D_h^n}{8^{n-1}K(b+\dfrac{a}{n})^n} = 579$
-     Thus, the flow is laminar and the friction factor is estimated as: f=1/578=0.0276 and

$$-\frac{\Delta P}{L} = \frac{2f\rho u^2}{D_h} = 1963 Pa / m$$

ii.   For a circular tube, the area of flow

-     For a circular pipe, a =0.25, b = 0.75, $D_h = D = 0.0678m$
-     Reynolds number, $\mathrm{Re}_g = \dfrac{\rho u^{2-n}D_h^n}{8^{n-1}K(b+\dfrac{a}{n})^n} = 1070$
-     The flow is laminar and the friction factor is estimated as: f=1/1070=0.01495 and

$$-\frac{\Delta P}{L} = \frac{2f\rho u^2}{D_h} = 689 Pa / m$$

iii. For a rectangular duct with H/W = 0.5, H =0.0425 m and W= 0.085 m (for the same area of flow), and from Table 6: a=0.244, b= 0.728

-     D_h = 4HW/2(H+W) = 0.0567m
-     Reynolds number, $\mathrm{Re}_g = \dfrac{\rho u^{2-n}D_h^n}{8^{n-1}K(b+\dfrac{a}{n})^n} = 960$

- f=1/960=0.0167 and $-\dfrac{\Delta P}{L} = \dfrac{2f\rho u^2}{D_h} = 919\,Pa\,/\,m$

iv. elliptical, $b'/a' = 0.5$.

- From Table 6 : a=0.2629, b= 0.7886
- $D_h = = 0.0607m$
- Reynolds number, $Re_g = 953$

- f=0.0168 and $-\dfrac{\Delta P}{L} = 864\,Pa\,/\,m$

9. References

Astarita, G., The Engineering Reality of the Yield Stress, *J. Rheol.* 34 (1990) 275

Barnes , H.A. and Walters , K. , The yield stress myth, *Rheol. Acta* 24 (1985) 323 .

Barnes , H.A. , Edwards , M.F. and Woodcock , L.V. , Applications of computer simulations to dense suspension rheology, *Chem. Eng. Sci.* 42 (1987) 591 .

Barnes , H.A. , Hutton , J.F. and Walters , K. , An Introduction to Rheology , *Elsevier*, Amsterdam (1989) .

Barnes , H.A., The yield stress - a review, *J. Non-Newt. Fluid Mech.* 81 (1999) 133.

Bird , R.B. , Useful non-Newtonian models, *Annu. Rev. Fluid Mech.* 8 (1976) 13.

Bird , R.B. , Dai , G.C. and Yarusso , B.J. , The rheology and flow of viscoplastic materials, *Rev. Chem. Eng.* 1 (1983) 1 .

Bird , R.B. , Armstrong , R.C. and Hassager , O. , Dynamics of polymeric liquids. Fluid dynamics, 2nd edition, Wiley , New York (1987) .

Cross , M.M. , Rheology of non-Newtonian fluids: A new flow equation for pseudoplastic systems, *J. Colloid Sci.* 20 (1965) 417.

Carreau , P.J. , Rheological equations from molecular network theories, *Trans. Soc. Rheol.* 16 (1972) 99.

Carreau , P.J. , Dekee , D. and Chhabra , R.P., Rheology of polymeric systems: Principles and applications , (1997)

Delplace , F. and Leuliet , J. C. , Generalized Reynolds number for the flow of power law fluids in cylindrical ducts of arbitrary cross-section, Chem. Eng. Journal, 56, 33-37., *Chem. Eng. J.* 56 (1995) 33.

Dodge, D. W. and Metzner, A. B., Turbulent flow of non-Newtonian systems, *AIChE Journal.* 5 (1959) 189-204.

Escudier , M.P. and Smith , S. , Non-Newtonian liquids through a square duct, *Proc. Royal Soc.* 457 (2001) 911 .

Evans , I. D. , On the nature of the yield stress, *J. Rheol.* 36 (1992) 1313 .

Goddard , J.D. and Bashir , Y. , Recent developments in structured continua II (Chapter 2) , Longman , London (1990) .

Griskey, R.G., Nechrebecki, D.G., Notheis , P.J. and Balmer , R.T., Rheological and pipeline flow dehavior of corn starch dispersions, *J. Rheol.* 29(1985) 349

Hartnett , J.P. and Kostic , M. , Heat transfer to Newtonian and non- Newtonian fluids in rectangular duct, *Adv. Heat Transf.* 19 (1989) 247.

Irvine , Jr. , T.F. , A generalized blasius equation for power law liquids, *Chem. Eng. Commun.* 65 (1988) 39.

Johnson , A.T. , Biological Process Engineering , Wiley , New York (1999).

Kozicki , W. , Chou , C.H. and Tiu , C. , Non-Newtonian flow in ducts of arbitrary cross-sectional shape, *Chem. Eng. Sci.* 21 (1966) 665 .

Kozicki , W. , and Tiu , C., Non-Newtonian flow through open channels, *Chem. Eng.* 45 (1967) 127 .

Kozicki , W. and Tiu , C. , Encyclopedia of Fluid Mechanics, Vol. 7, Gulf, Houston , p. 199 (1988).

Kostic , M. and Hartnett , J.P. , Predicting turbulent friction factors of non- Newtonian fluids in noncircular ducts, *Int. Comm. Heat Mass Transf.* 11 (1984) 345 .

Metzner , A.B. and Whitlock , M. , Flow behavior of concentrated (dilatant) suspensions, *Trans. Soc. Rheol.* 2 (1958) 239 .

Metzner, A.B. and Reed, J.C., Flow of non-Newtonian fluids correlation of the laminar, transition, and turbulent-flow regimes, *AIChE Journal.* 1 (1955) 434-40.

Miller , C. , Predicting non-Newtonian flow behaviour in ducts of unusual cross-sections *Ind. Eng. Chem. Fundam.* 11 (1972) 524 .

Mitsuishi , N. and Aoyagi , Y. , Non-Newtonian flow in non- circular ducts, *Chem. Eng. Sci .* 24 (1969) 309 .

Mitsuishi , N. and Aoyagi , Y. , Non-Newtonian fluid flow in an eccentric annulus, *J. Chem. Eng. Jpn.* 6 (1973) 402 .

Ryan, N.W. and Johnson, M.M., Transition from laminar to turbulent flow in pipes, *AIChE Journal.* 5 (1959) 433.

Schechter, R.S. , On the steady flow of a non-Newtonian fluid in cylinder ducts, *AIChE J.* 7 (1961) 445 .

Schurz , J. , The yield stress - an empirical reality, *Rheol. Acta.* 29 (1990) 170 .

Uhlherr , P.H.T. , Guo , J. , Zhang , X-M. , Zhou , J.Z and Tiu , C. , The shear-induced solid-liquid transition in yield stress materials with chemically different structures, *J. Non-Newt. Fluid Mech.* 125 (2005) 101 .

Wheeler , J.A. and Wissler , E.H. , The Friction factor-Reynolds number relation for a steady flow of pseudoplastic fluids through rectangular ducts, *AIChE J.* 11 (1965) 207.

5

Noise and Vibration
in Complex Hydraulic Tubing Systems

Chuan-Chiang Chen
Mechanical Engineering Department,
California State Polytechnic University Pomona,
USA

1. Introduction

In hydraulic systems, pumps are the major source of noise and vibration. It generates flow ripples which interact with other hydraulic components, such as transmission lines and valves to create harmonic pressure waves, *i.e.*, fluid-borne noise (FBN). Fig. 1 shows a typical oscillating pressure measured at the outlet of a ten-vane pump running at 1500 rpm. Fig. 2 gives the frequency spectrum for the pressure signal which contains harmonic components of the fundamental frequency, 25 Hz, which correlates with the pump operating speed. The largest peak is at 250 Hz, which corresponds to the shaft speed times the number of the pumping elements (10 vanes in this case). The FBN propagates along as well as interacts with the tubing and other components to result in airborne noise (ABN) and structure-borne noise (SBN, *i.e.*, structural vibration). These noises can become excessive, and lead to damage the tubing system and other components. Therefore, to study the pressure wave propagation in the hydraulic tubing system, it is important to take the fluid-structure interaction into account to further the understanding of noise transmission mechanism.

Fluid-structure interaction can be divided into three categories: junction coupling, Poisson coupling, and Bourdon coupling. Junction coupling occurs at discontinuities, such as bends and tees, where the pressure interacts with the structure to cause structural vibration. In unsteady flow, the pressure varies along the tube. Differences in pressure exert axial and transverse forces during power transmission at bends and other locations where the diametrical geometry changes. Moreover, the pressure is related to the longitudinal stresses in the pipe because of the radial contraction or expansion via Poisson coupling (Hatfield & Davidson, 1983). Furthermore, the cross-sectional shape of the line in a bend is not circular because of action by the bending forces. This effect, known as the Bourdon effect (Tentarelli, 1990), influences the structural modes at low frequencies.

Several approaches have been used (To & Kaladi, 1985; Everstine 1986; Nanayakkara & Perreia, 1986), such as the transfer matrix and finite element (FEM) methods, to model the fluid-structural coupling. In this study, the transfer matrix method (TMM) is used because of its simplicity. Even though FEM may offer better accuracy, it is more complicated and time-consuming than TMM.

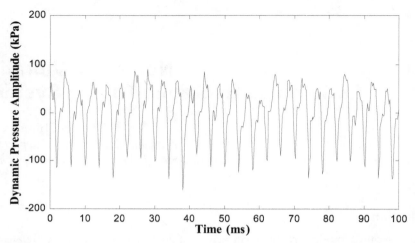

Fig. 1. Pressure waveform measured at the outlet of a ten-vane power steering pump running at 1500 rpm. The periodic waveform is generated by the rotating elements of the pumping mechanism.

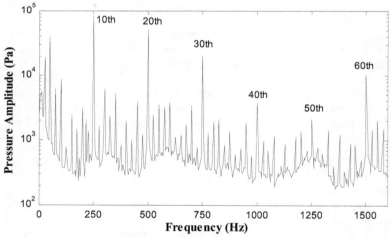

Fig. 2. Frequency spectrum of the pressure signal shown in Fig. 1 Pump speed: 1500 rpm; number of pumping elements: 10; fundamental pump rotational frequency: 25 Hz

Davidson and Smith (1969) first studied fluid-structure interactions using the TMM and verified their model with their own experimental data. Their data were used widely by subsequent researchers (Davidson & Samsury, 1972; Hatfield & Davidson, 1983) to verify analytical models which did not include viscosity. Hatfield *et al.* (1982) applied the component synthesis method in the frequency domain. In their method, fluid-structure interaction was included in terms of junction coupling. Their simulation predictions were validated with Davidson and Smith's (1969) experimental data. Bundy *et al.* [9] introduced structural damping which was neglected by other researchers in previous experimental and theoretical investigations.

Brown and Tentarelli (1988) arranged the 14×14 transfer matrices for n segments and then assembled them into a global 14(n-1)×14(n-1) sparse matrix. This approach was beneficial because, by solving the linear equations, the state variables at every point were obtained. Their algorithm also avoided round-off error at higher frequencies. Fluid friction was not considered in their analysis. Chen (1992), and Chen and Hastings (1992; 1994) considered both the fluid-structure interaction caused by discontinuities and the viscosity of the fluid in a distributed parameter, transfer matrix model of the transmission line in an automotive power steering system.

Most researchers verified their models with a simplified experimental system; for example, L-tube or U-tube systems. Until now, the system model has not been verified in a complex tubing system. In this book, a transfer matrix system model incorporating the acoustic characteristics of termination is developed to predict the fluidborne noise in a complex three-dimensional tubing system. The results show good agreements between simulated and experimental data.

2. Analysis

2.1 Axial motion

For a three-dimensional tubing system, fluid–structural coupling must be considered because tubing discontinuities, such as bends, cause unbalanced forces to act on both the tubing and fluid. Fig. 3 displays the coordinate system and state variables in a straight tube segment used in the following analysis.

Assuming axisymmetric, two-dimensional, laminar, viscous, compressible flow and negligible temperature variation (*i.e.*, constant fluid viscosity), the linearized Navier–Stokes equations reduce to (Chen, 2001):

$$\frac{\partial v_z}{\partial t} = -\frac{1}{\rho_f}\frac{\partial p}{\partial z} + \upsilon\left[\frac{\partial^2 v_z}{\partial r^2} + \frac{1}{r}\frac{\partial v_z}{\partial r}\right] \qquad (1)$$

where v_z, v_r, and p denote the deviation of axial velocity, radial velocity, and pressure from the steady state, respectively.

Combining the continuity equation and equation of state for a liquid, gives:

$$\frac{1}{\beta}\frac{\partial p}{\partial t} + \frac{\partial v_r}{\partial r} + \frac{v_r}{r} + \frac{\partial v_z}{\partial z} = 0 \qquad (2)$$

where β is the fluid bulk modulus.

By averaging v_z over the cross section, applying the boundary condition at the inner radius of the tubing, $u_f = u_z$, and transforming to the Laplace domain, the following equation is obtained:

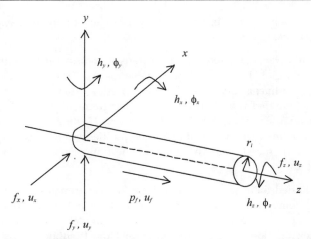

Fig. 3. Hydraulic line coordinate system and state variables: u is translational displacement of tubing; ϕ, angular displacement of tubing; f, force acting on the tubing; h, moment acting on the tubing; p_f, fluid pressure; u_f, fluid displacement; and subscripts x, y, and z the axes for the Cartesian coordinates (adapted from Chen, 1992).

$$\frac{\partial P}{\partial z} = \frac{\rho_f s^2}{\Omega(s)} U_f - \rho_f s^2 \left[\frac{1+\Omega(s)}{\Omega(s)} \right] U_z \qquad (3)$$

where $\Omega(s) = \left[\dfrac{2J_1\left(jr_i\sqrt{s/\upsilon}\right)}{jr_i\sqrt{s/\upsilon}J_i\left(jr_i\sqrt{s/\upsilon}\right)} - 1 \right]$, ρ_f is the fluid density, and J_0 and J_1 are the zero– and

first–order Bessel functions of the first kind, respectively, and s denotes the Laplace transformation.

Applying Newton's second law to the tubing wall, yields:

$$F_\tau + \frac{\partial F_z}{\partial z} = \rho A s^2 U_z \qquad (4)$$

where F_τ is the friction force per unit length acting on the inner tubing wall, A is the cross-sectional area of the tubing, and ρ is the density of the tubing.

Applying Newton's second law to the fluid gives:

$$-A_f \frac{\partial P}{\partial z} - F_\tau = \rho_f A_f s^2 U_f \qquad (5)$$

where A_f is the cross-sectional area of the fluid.

Combining Equations (4) and (5), gives:

$$\frac{\partial F_z}{\partial z} - A_f \frac{\partial P}{\partial z} = \rho A s^2 U_z + \rho_f A_f s^2 U_f \qquad (6)$$

Substituting Equation (3) into Equation (6), and rearranging Equation (6) yields:

$$\frac{\partial F_z}{\partial z} = \left\{ \rho A - \rho_f A_f \left[1 + \frac{1}{\Omega(s)} \right] \right\} s^2 U_z + \rho_f A_f \left[1 + \frac{1}{\Omega(s)} \right] s^2 U_f \qquad (7)$$

2.1.2 Poisson effect

The Poisson effect, longitudinal motion resulting in radial strain of the tubing or vice versa, was not included in previous work (Chen, 1992). The axial strain of the tubing, ε_z, in a cylindrical coordinates is written as:

$$\varepsilon_z = \frac{\partial u_z}{\partial z} = \frac{1}{E}(\sigma_z - \mu\sigma_r - \mu\sigma_\theta) \qquad (8)$$

where σ is the stress, E and μ are the elastic modulus and Poisson's ratio for the tubing material, respectively, and subscripts z, r and θ are the cylindrical coordinates.

For thick-walled tubing, the radial and tangential stresses can be represented as:

$$\sigma_r + \sigma_\theta = \frac{2pr_i^2}{r_o^2 - r_i^2} \qquad (9)$$

Combining Equations (8) and (9), gives:

$$\frac{\partial u_z}{\partial z} = \frac{1}{EA} f_z - \frac{2}{E} \frac{r_i^2 \mu}{(r_o^2 - r_i^2)} p \qquad (10)$$

For conservation of mass to hold, the axial change in volume of a fluid element results from pressure and expansion of the tubing. Radial expansion of the tubing is caused by pressure, and axial motion of the tubing results from Poisson coupling:

$$\frac{\partial u_f}{\partial z} = \frac{2\mu}{EA} f_z - \frac{1}{\beta_e} p \qquad (11)$$

where β_e is an effective fluid bulk modulus that accounts for compliance of the tubing wall. When the Poisson effect is neglected, Equations (10) and (11) reduce to equations for longitudinal motion of a bar.

2.1.3 Bourdon coupling

The Bourdon effect occurs at bends where the fluid-filled tubing cross-section is ovalized. Bending of the tubing results in a change of cross-sectional area and thus fluid motion. The fluid pressure gradient in the bend produces a bending moment in the tubing, and the balancing bending moment in the tubing then displaces the fluid. For curved tubing, the Bourdon coupling is described by Reissner et al. (1952) and Tentarelli (1990):

$$\frac{\partial u_z}{\partial z} = A_{11}p + A_{12}h_y \qquad (12)$$

$$\frac{\partial \phi_y}{\partial z} = A_{21}p + A_{22}h_y \tag{13}$$

where
$$A_{11} = \frac{1}{R_v}\left(1-\frac{b^2}{a^2}\right)\left[1-\frac{2R_v(r_o-r_i)}{ab\sqrt{12(1-v^2)}}\right]\frac{b^2-a^2}{Ea(r_o-r_i)^2}\left[\sqrt{3(1-v^2)}-\frac{R_v(r_o-r_i)}{ab}\right],$$

$$A_{22} = \frac{a\sqrt{12(1-v^2)}}{2\pi Eb^2 R_v(r_o-r_i)^2}, \qquad\qquad A_{12} = \frac{-1}{\pi R_v ab}\frac{b^2-a^2}{Ea(r_o-r_i)^2}\left[\sqrt{3(1-v^2)}-\frac{R_v(r_o-r_i)}{ab}\right],$$

$$A_{21} = \frac{b^2-a^2}{Ea(r_o-r_i)^2}\left[\sqrt{3(1-v^2)}-\frac{R_v(r_o-r_i)}{ab}\right], \quad R_v \text{ is the radius of curvature of the bend, and } a$$

and b are the major and minor axes of the elliptical cross section, respectively.

Equations (12) and (13) reduce to the common flexural motion equations for $a=b$ (i.e., circular cross-section). When there is no fluid pressure present, A_{22} can be approximated as a flexural stiffness with a correction factor to account for the ovalization effect. The effect produces a reduction in stiffness at bends in the transmission line.

Several straight short-length segments are used to model the bends and twists in the three-dimensional tubing line. To account for the ovalization effect, a correction factor is used to adjust the flexural stiffness for the curved line. The correction factor (κ) for the flexural stiffness is formulated as Vigness (1943):

$$\kappa = \frac{1+12\ [4(r_o-r_i)R_v/(r_o+r_i)^2]}{10+12\ [4(r_o-r_i)R_v/(r_o+r_i)^2]} \tag{14}$$

The product of κ and a flexural stiffness can be shown to be a simplified form of A_{22} (Reissner et al., 1956).

2.2 Flexural and torsional motion

Rearranging Equations (3), (7), (10) and (11), and considering the flexural motions in the x-z and y-z planes, and torsion about the z-axis in Laplace domain, four groups of linear, first-order differential equations are obtained (Chen 2001):

$$\frac{\partial}{\partial z}\begin{bmatrix} P \\ F_z \\ U_f \\ U_z \end{bmatrix} = -\begin{bmatrix} 0 & 0 & \frac{-\rho_f s^2}{\Omega(s)} & \rho_f s^2\left[\frac{1+\Omega(s)}{\Omega(s)}\right] \\ 0 & 0 & -\rho_f A_f s^2\left[\frac{1+\Omega(s)}{\Omega(s)}\right] & -s^2\left\{\rho A - \rho_f A_f\left[\frac{1+\Omega(s)}{\Omega(s)}\right]\right\} \\ \frac{1}{\beta_e} & \frac{-2\mu}{EA} & 0 & 0 \\ \frac{2\mu r_i^2}{E(r_o^2-r_i^2)} & \frac{-1}{EA} & 0 & 0 \end{bmatrix}\begin{bmatrix} P \\ F_z \\ U_f \\ U_z \end{bmatrix} \tag{15}$$

$$\frac{\partial}{\partial z}\begin{bmatrix} U_x \\ H_y \\ F_x \\ \Phi_y \end{bmatrix} = -\begin{bmatrix} 0 & 0 & -1/GA & -1 \\ 0 & 0 & 1 & -\left(\rho I + \rho_f I_f\right)s^2 \\ -\left(\rho A + \rho_f A_f\right)s^2 & 0 & 0 & 0 \\ 0 & -1/EI & 0 & 0 \end{bmatrix}\begin{bmatrix} U_x \\ H_y \\ F_x \\ \Phi_y \end{bmatrix} \tag{16}$$

$$\frac{\partial}{\partial z}\begin{bmatrix} U_y \\ H_x \\ F_y \\ \Phi_x \end{bmatrix} = -\begin{bmatrix} 0 & 0 & -1/GA & 1 \\ 0 & 0 & -1 & -\left(\rho I + \rho_f I_f\right)s^2 \\ -\left(\rho A + \rho_f A_f\right)s^2 & 0 & 0 & 0 \\ 0 & -1/EI & 0 & 0 \end{bmatrix}\begin{bmatrix} U_y \\ H_x \\ F_y \\ \Phi_x \end{bmatrix} \tag{17}$$

$$\frac{\partial}{\partial z}\begin{bmatrix} H_z \\ \Phi_z \end{bmatrix} = -\begin{bmatrix} 0 & -\rho J s^2 \\ -1/GJ & 0 \end{bmatrix}\begin{bmatrix} H_z \\ \Phi_z \end{bmatrix} \tag{18}$$

Equations (15) – (18) can be represented in the following form:

$$\frac{\partial}{\partial z}[S_k] = -[A_k]\,[S_k]\ ; k = 1,4 \tag{19}$$

where $[A_k]$ is coefficient matrix, $S_1 = \begin{bmatrix} P & F_z & U_f & U_z \end{bmatrix}^T$, $S_2 = \begin{bmatrix} U_x & H_y & F_x & \Phi_y \end{bmatrix}^T$, $S_3 = \begin{bmatrix} U_y & H_x & F_y & \Phi_x \end{bmatrix}^T$ and $S_4 = \begin{bmatrix} H_z & \Phi_z \end{bmatrix}^T$.

Solving Equation (16) by employing boundary conditions at the inlet ($z = 0$) of each section yields:

$$[S_k]_{z=L} = e^{-[A_k]L}[S_k]_{z=0} \tag{20}$$

where $[S_k]_{z=0}$ is the substate vector at the inlet.

Relating the two end conditions for a given section i, yields the 14×14 field transfer matrix $[T]_i$:

$$[S]_{i+1} = [T]_i\,[S]_i \tag{21}$$

A three-dimensional tubing system can be treated as a combination of short straight lines with different orientations resulting in coupling of the fluid pressure, and forces and moments in the tubing wall. Each section of tubing is modeled by a 14×14 transfer matrix with state variable vectors. Details on the assembly of the 14×14 matrix can be found in Chen [11]. Each bend is broken into three straight-line segments. For these segments, the correction factor, κ, is used to include the Bourdon effect by replacing the flexural stiffness EI with κEI in Equation (14).

A transformation matrix [R] transfers the force and displacement from one section to another, couples structural vibration and fluid pressure waves at points of discontinuity, and transforms the coordinate system from one section to the next. Force and moment equilibrium, conservation of mass flow, and structural continuity are considered when deriving the transformation matrix (Chen, 1992). Finally, the relationship between one end of the system and the other is obtained by multiplying $[R]$ and $[T]$ for each line section:

$$[S]_{n+1} = [R]_n [T]_n \cdots\cdots [R]_1 [T]_1 [S]_1 \tag{22}$$

2.3 Implementation of the matrix partitioning algorithm

The transfer matrix method solves the equations of motion step by step and determines the unknown variables (translational displacement, angular displacement, force and moment) simultaneously in the solution process. Because of the transfer matrix chain multiplication, as shown in Equation (22), numerical errors occur and build up as the multiplicative process progresses. In this study, matrix partitioning was applied to the system of equations to eliminate the long chain of matrix multiplication.

In most tubing systems, the boundary conditions at each end are defined because the tubing is attached to the pump outlet and the rotary valve inlet. Therefore, to reduce numerical error, matrix partitioning originally developed by (Clark, 1956) was used. With known boundary conditions at the ends, the state variables are re-arranged as follows:

$$[S]_1^* = \begin{Bmatrix} S_1{}^a \\ S_1{}^b \end{Bmatrix} \tag{23}$$

$$[S]_{n+1}^* = \begin{Bmatrix} S_{n+1}{}^a \\ S_{n+1}{}^b \end{Bmatrix} \tag{24}$$

where $S_1{}^a$ and $S_{n+1}{}^a$ are the known state variables, and $S_1{}^b$ and $S_{n+1}{}^b$ are the unknown variables.

By using the matrices $[MR]_1$ and $[MR]_{n+1}$, the following equations are obtained:

$$[S]_1^* = [MR]_1 [S]_1 \tag{25}$$

$$[S]_{n+1}^* = [MR]_{n+1} [S]_{n+1} \tag{26}$$

$$[S]_1 = [MR]_1^{-1} [S]_1^* \tag{27}$$

The relationship between one end and the other for the first element is:

$$[S]_2 = [TR]_{21} [S]_1 \tag{28}$$

where $[TR]_{21} = [R]_1 [T]_1$.

Combining Equations (27) and (28) and arranging the unknown variables on the left side, yields:

$$\begin{bmatrix} \underset{1:14,\,b+1:14}{TR_{21}^*} & -\underset{1:14,\,1:14}{I} \\ & \end{bmatrix}_{14\times(a+14)} \begin{bmatrix} S_1^{\ b} \\ S_2 \end{bmatrix}_{(a+14)\times 1} = -[TR]_{21}^* \underset{14\times 14}{} \begin{bmatrix} S_1^{\ a} \\ 0 \end{bmatrix}_{14\times 1} \tag{29}$$

where $[TR]_{21}^* = [TR]_{21}[MR]_1^{-1}$.

Similarly, the equation of the last section of tubing can be written as:

$$\begin{bmatrix} 0 & \\ \underset{1:14,1:14}{TR_{n+1,n}^*} & \underset{1:d,1:d}{-I} \end{bmatrix}_{14\times(d+14)} \begin{bmatrix} S_n \\ S_{n+1}^{\ b} \end{bmatrix}_{(d+14)\times 1} = \begin{bmatrix} S_{n+1}^{\ a} \\ 0 \end{bmatrix}_{14\times 1} \tag{30}$$

where $[TR]_{n+1,n}^* = [MR]_{n+1}[TR]_{n+1,n}$ and $[TR]_{n+1,\,n} = [R]_n[T]_n$.

By rearranging the equations for all tubing sections, the global matrix is obtained:

$$\begin{bmatrix} [TR]_{21}^* & -I & & & \\ & [TR]_{32} & -I & & \\ & & \ddots & & \\ & & & [TR]_{n,n-1} & -I \\ & & & & [TR]_{n+1,n}^* & [-I] \end{bmatrix} \begin{bmatrix} S_1^{\ b} \\ S_2 \\ \vdots \\ S_n \\ S_{n+1}^{\ d} \end{bmatrix} = \begin{bmatrix} -[TR]_{21}^* \begin{bmatrix} S_1^{\ a} \\ 0 \end{bmatrix} \\ 0 \\ \vdots \\ 0 \\ \begin{bmatrix} S_{n+1}^{\ c} \\ 0 \end{bmatrix} \end{bmatrix} \tag{31}$$

To solve for the unknown state variables in Equation (31), MATLAB® command "\", which solves the system of linear equations by Gaussian elimination, was used in the simulation.

2.4 Acoustic impedance of hydraulic system components

Impedance characteristics of hydraulic components have an important effect on pressure pulsations in hydraulic circuits. These pressure oscillations lead to vibrations and are a source of noise. By using plane wave propagation theory, impedances can be estimated using the two-microphone technique (ASTM E 1050-90, and ASTM C 384-108a).

Fig. 4 displays an acoustic impedance representation for the hydraulic circuit Five parameters can be used to define this system: the source impedance (Z_s), source flow ripple (Q_s), line impedance (Z_c), line propagation constant (Γ), and termination impedance (Z_t). The source pressure (P_s) is derived from Z_s.

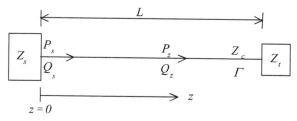

Fig. 4. Acoustic representation of a hydraulic circuit

The pressure (P_z) and volumetric flow velocity (Q_z) at any harmonic frequency at a distance z along the line are:

$$P_z = P_i e^{-\Gamma z} + P_r e^{\Gamma z} \tag{32}$$

$$Q_z = \frac{1}{Z_c}\left(P_i e^{-\Gamma z} - P_r e^{\Gamma z}\right) \tag{33}$$

where P_i and P_r are the complex incident and reflected pressures, respectively.

The termination reflection coefficient, C_r, is defined by the ratio of the reflected pressure to the incident pressure:

$$C_r = \frac{P_r}{P_i} \tag{34}$$

The pressures at locations z_1 and z_2 are:

$$P_{z1} = P_i e^{-\Gamma z_1} + P_r e^{\Gamma z_1} \tag{35}$$

$$P_{z2} = P_i e^{-\Gamma z_2} + P_r e^{\Gamma z_2} \tag{36}$$

Solving Equations (35) and (36) for P_i and P_r, and substituting into Equation (34), the reflection coefficient is obtained:

$$R = \frac{-\left(P_{z1}e^{-\Gamma z_2} - P_{z2}e^{-\Gamma z_1}\right)}{P_{z1}e^{\Gamma z_2} - P_{z2}e^{\Gamma z_1}} \tag{37}$$

If the impedance of the termination is Z_t, applying Equations (32) and (33) at the boundary $z' = 0$ (*i.e.*, $z = L$) gives:

$$\left(P_i + P_r\right)_{z'=0} = \left(P_t\right)_{z'=0} \tag{38}$$

$$\left[\left(P_i - P_r\right)/Z_c\right]_{z'=0} = \left(P_t/Z_t\right)_{z'=0} \tag{39}$$

By rearranging the above two equations and combining with Equation (34), the termination impedance can be represented in terms of Z_c and C_r:

$$Z_t = \left(\frac{1+C_r}{1-C_r}\right)Z_c \tag{40}$$

Z_c is the characteristic impedance in the tubing given by Chen and Hastings (1992):

$$Z_c = \frac{\rho_f \, c}{\pi \, r_i^2}\left[1 - \frac{2}{jr_i\sqrt{j\omega/\upsilon}}\frac{J_1(jr_i\sqrt{j\omega/\upsilon})}{J_0(jr_i\sqrt{j\omega/\upsilon})}\right]^{-\frac{1}{2}} \tag{41}$$

where c is the sound speed; ω is the angular frequency; r_i is the inner radius of transmission line; v is the density and kinematic viscosity of the fluid, respectively; and J_0 and J_1 denote the zero- and first-order Bessel functions of the first kind, respectively.

The measuring pressure signals before the valve, the termination impedance can be readily determined by Equation (37) and (40). Fig. 5 displays the estimated impedance of the rotary valve in power steering system at various opening positions. The data show that modeling this valve as a pure resistance is not appropriate as a strong reactive component of the impedance is apparent.

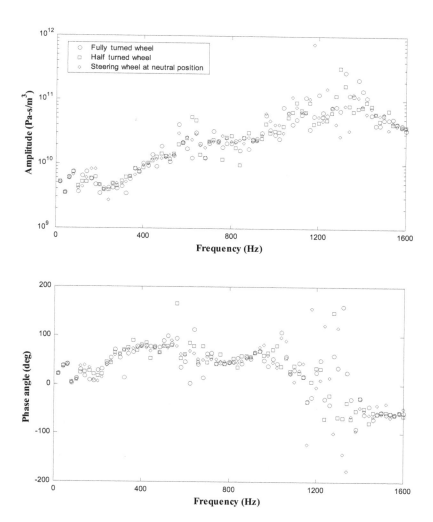

Fig. 5. Amplitude and phase angle of the valve impedance

3. Experimental results

An automotive hydraulic power steering tubing system was tested in this research. Detailed layout of this three-dimensional tubing transmission line was provided by the manufacturer. Since this study addresses pump induced noise, a system with a pump source was set up to verify the transfer matrix model for the tubing system. Fig. 6 illustrates the system layout. This includes the power steering pump, hydraulic transmission lines, a rack and pinion unit, steering wheel and column, and rotary valve. The steering pump is driven by an electric motor through a belt. A variable speed, AC controller is used to control the electric motor and vary the speed of pump. In this setup, a water-cooling system using a coil heat exchanger is used. Water from the building supply circulates through the coil heat exchanger connected to the return line and then flows into a drain.

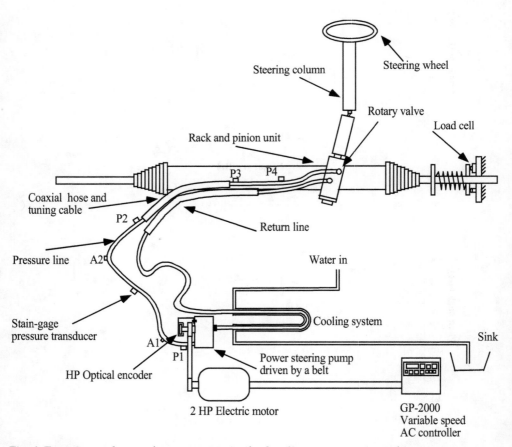

Fig. 6. Experimental setup for an automotive hydraulic power steering tubing system

Four piezoelectric pressure transducers are used to measure the dynamic pressure in this system. The first pressure transducer (P1) is placed at the outlet of the pump to measure the source of pressure disturbance. The fourth one (P4) is located at the inlet of the rotary valve so that the amplitude ratio of outlet pressure to inlet pressure (P4/P1) can be measured and compared to the model prediction. The pressure signals are connected to the Kistler Charge amplifier and then to a HP3566A 8-channel analyzer. Data are saved in a computer and retrieved later for further analysis.

Because the focus of this study is to investigate the fluidborne noise propagation in the tubing system and interaction with the tubing structure, the pressure frequency response of the tubing transmission line is investigated. To correlate the transfer matrix model with better accuracy, the sound speed in steel tubing and damping factor are experimentally estimated (Chen, 2001). The sound speed was optimized to be 1374 m/s. The frequency-dependent damping in the system was estimated based on the Half-Power method. Figs. 7 and 8 display the frequency response for the outlet pressure of the pressure side transmission line (P4) with different valve opening due to the steering wheel positions for an all steel tubing system. The model prediction and experimental data match very well. Good agreement was obtained.

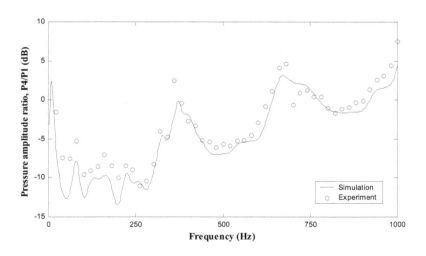

Fig. 7. Pressure response of an all steel tubing system with a fully turned steering wheel.

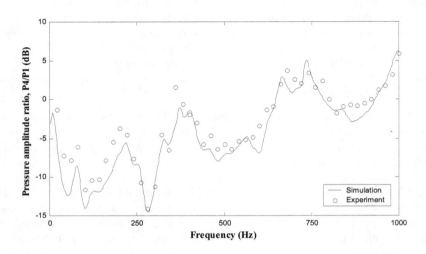

Fig. 8. Pressure response of an all steel tubing system with a steering wheel at neutral position

4. Conclusions

A distributed-parameter transfer-matrix model is developed to predict the fluidborne noise in a complex tubing system. This study provides a systematic approach to predict the pump-induced fluidborne noise by incorporating the experimentally determined acoustic characteristics of valve termination. The developed model was supported by experimental measurement with good agreements. Inclusion of Poisson and Bourdon effects in the model provide better predictions. Furthermore, the transfer matrix-partitioning algorithm presented here not only can reduce truncation error but also be more efficient in comparison with the matrix chain multiplication. It is also noted that the damping of the tubing system needs to be included to better predict the peak amplitude. The mathematical model presented can be applied to the analysis of noise in other hydraulic systems, such as those used in air conditioners and power plants. However, to fully characterize the noise propagation/transmission in the tubing system, SBN (not presented here, but can also be predicted by the developed model) should also be investigated because of fluid-structure interaction.

5. References

Brown F.T. & Tentarelli S.C. (1988). Analysis of Noise and Vibration in Complex Tubing System with Fluid-Filled Interactions. 43rd National Conference on Fluid Power, 1988, pp. 139-149

Bundy, D.D.; Wiggert, D.C. & Hatfield F.J. (1991). The Influence of Structural Damping on Internal Pressure during a Transient Flow, ASME Journal of Fluid Engineers, Vol. 113, pp. 424-579

Chen C.-C. (1992). A Theoretical Analysis of Noise Reduction in Automotive Power Steering Transmission Lines. M.S. Thesis, 1992, The Ohio State University, Columbus, Ohio

Chen C.-C. & Hastings M.C. (1992). Noise reduction in Power Steering Transmission Lines. Proceedings of the International Congress on Noise Control Engineering, 1992, Vol. 1, pp. 67-72

Chen C.-C. & Hastings M.C. (1994). Half-wavelength Tuning Cable for Passive Noise Control in Automotive Power Steering Systems. Active Control of Vibration and Noise (K. W. Wang et al., eds.), 1994, ASME DE-75, pp. 355-361

Chen C.-C. (2001). An Investigation of Noise and Vibration in an Automotive Power Steering System. Ph.D. Dissertation, 2001, The Ohio State University, Columbus, Ohio

Clark RA. Torsional Wave Propagation in Hollow Cylindrical Bars. Journal of Acoustical Society of America, 1956; 28 (6), pp. 1163-1165

Davidson L.C. & Smith, J.E (1969). Liquid-Structure Coupling in Curved Pipes. The Shock and Vibration Bulletin, 1969, Vol. 40, No. 4, pp. 197-207

Davidson L.C. & Samsury D.R. (1972). Liquid-Structure Coupling in Curved Pipes – II. The Shock and Vibration Bulletin, 1972, Vol. 43. No. 1, pp. 123-135

Everstine G.C. (1986). Dynamic Analysis of Fluid-Filled Piping Systems Using Finite Element Techniques. Journal of Pressure Vessel Technology, 1986, Vol. 10, pp. 57-61

Hatfield F.J. & Davidson L.C. (1983). Experimental Validation of the Component Synthesis Method for Prediction Vibration of Liquid-Filled Piping. The Shock and Vibration Bulletin, 1983, Vol. 53, No. 2, pp. 1-10

Hatfield, F.J. , Wiggert D.C. & Otwell R.S. (1982). Fluid Structure Interaction in Piping by Component Synthesis. ASME Journal of Fluid Engineers, 1982, Vol. 104, pp. 318-325

Nanayakkara S. & Perreia, N.D. (1986). Wave Propagation and Attenuation in Piping Systems. Journal of Vibration, Acoustics, Stress, and Reliability in Design, 1986, Vol. 108, pp. 441-446

Reissner E., Clark R.A. & Gilroy R.I. (1952). Stresses and Deformations of Torsional Shells of an Elliptical Cross Section with Applications to the Problems of Bending of Curved Tubes and the Bourdon Gage. Transaction of ASME, Journal of Applied Mechanics, 1952, pp.37-48

Tentarelli S.C. (1990). Propagation of Noise and Vibration in Complex Hydraulic Tubing System. Ph.D. Dissertation, 1990, Lehigh University, Bethlehem, Pennsylvania

To C.W.S. & Kaladi V. (1985). Vibration of Piping Systems Containing a Moving Medium. Transaction of ASME, Journal of Pressure Vessel Technology, 1985. Vol. 107, pp. 344-349

Vigness I. (1943). Elastic Properties of Curved Tubes. Transaction of ASME, 1943, Vol.65, pp.105-117

6

Analysis Precision Machining Process Using Finite Element Method

Xuesong Han

School of Mechanical Engineering, Tianjin University,
P.R. China

1 Introduction

Machining is the process of removing the material in the form of chips by means of wedge shaped tool[1]. The need to manufacture high precision items and to machine difficult-to-cut materials led to the development of the newer machining processes. The dimensional tolerance achieved by precision machining technology is on the order of 0.01 μm and the surface roughness is on the order of 1 nm. The dimensions of the parts or elements of the parts produced may be as small as 1 μm, and the resolution and the repeatability of the machine used must be of the order of 0.01 μm (10 nm). The accuracy targets for ultra-precision component cannot be achieved by a simple extension of conventional machining processes and techniques. They are called precision machining processes, notwithstanding that the definition of conventional and traditional changes with time. Unlike conventional machining processes, precision machining processes are not based on the removing the metal in the form of chips using a wedge shaped tool. There are a variety of ways by which the material may be removed in precision machining processes. Some of them are abrasion by abrasive particles, impact of water, thermal action, chemical action and so on.

When metal is removed by machining there is substantial increase in the specific energy required with decrease in chip size. It is generally believed this is due to the fact that all metals contain defects (grain boundaries, missing and impurity atoms, etc.), and when the size of the material removed decreases, the probability of encountering a stress-reducing defect decreases. Since the shear stress and strain in metal cutting is unusually high, discontinuous microcracks usually form on the metal-cutting shear plane. If the material being cut is very brittle, or the compressive stress on the shear plane is relatively low, microcracks grow into gross cracks giving rise to discontinuous chip formation[2]. When discontinuous microcracks form on the shear plane they weld and reform as strain proceeds, thus joining the transport of dislocations in accounting for the total slip of the shear plane. In the presence of a contaminant, the rewelding of microcracks decreases, resulting in decrease in the cutting force required for chip formation. Owing to the complexity of elastic-plastic deformation at nanometer scale, the world wide convinced precision materials removal theory is not built up until now.

There are two basic approaches to the analysis of metal cutting process, namely, the analytical and the numerical method. As the complexity associate with the precision machining process, which involve high strains, strain rates, size effects and temperature, various simplifications and idealizations are necessary and therefore important machining features such as the strain hardening, strain rate sensitivity, temperature dependence, chip formation and the chip-tool interface behaviors are not fully accounted for by the analytical methods. Experimental studies on precision machining are expensive and time consuming. Moreover, their results are valid only for the experimental conditions used and depend greatly on the accuracy of calibration of the experimental equipment and apparatus used. Advanced numerical techniques such as Finite Element Method is a potential alternative for solving precision machining problems.

Finite Element Method (FEM) which is originated from continuum mechanics, has already been justified as successful method in analyzing complicated engineering problem[3-8]. There are many advantages of using FEM to investigate machining: multi-physical machining variables output can be acquired (cutting force, chip geometry, stress and temperature distributions), improving precision and the efficiency comparing with Try-Out-Method and so on. In the last three decades, FEM has been progressively applied to metal cutting simulations. Starting with two-dimension simulations of the orthogonal cutting more than two decades ago, researches progressed to three-dimensional FEM models of the oblique cutting, which capable of simulating metal cutting processes such as turning and milling. Increased computation power and the development of robust calculation algorithms (thus widely availability of FEM programs) are two major contributors to this progress. Unfortunately, this progress was not accompanied by new developments in precision machining theory so the age-old problems such as the chip formation mechanism and tribology of the contact surfaces are not modeled properly. Further, even at a moderate cutting speed, the strain rates are quite high, almost of the order of 10^4 per second and the temperature rise is also quite large. As a result, the visco-plasticity and temperature-softening effects become more important compared to strain-hardening. Therefore, the material properties associated with these two effects should be known for a range of strain rates and temperatures occurring in typical machining processes. Additionally, to incorporate the temperature rise in the analysis, one needs to solve the heat transfer equation governing the temperature field in conjunction with the usual three equations governing the deformation field. For plastic deformation, these equations are coupled, and hence difficult to solve.

In material removal processes at the precision scale, the undeformed chip thickness can be on the order of a few microns or less, and can approach the nanoscale in some cases. At these length scales, the surface, subsurface, and edge condition of machined features and the fundamental mechanism for chip formation are much more intimately affected by the material properties and microstructure of the workpiece material, such as ductile/brittle behavior, crystallographic orientation of the material at the tool/chip interface, and micro-topographical features such as voids, secondary phases, and interstitial particulates. Characterizing the surface, subsurface, and edge condition of machined features at the precision scale in the FEM analysis are of increasing importance for understanding, and controlling the manufacturing process. There are still many challenges in the investigation of precision machining by means of FEM.

As mentioned above, this chapter will give some key factors on numerical modeling of precision machining and current advancements.

2. The flow stress characteristics of the workpiece materials

The flow stress characteristics are an important issue in the numerical analysis which is directly affects the loads and stresses in the precision machining. The flow stress is generally considered as function of strain, strain rate and temperature. Many research works justify that the influence of strain rate on flow stress become more important when the temperature becomes higher. It is important to build the appropriate flow stress models fit for different working conditions.

Accuracy and reliability of the predictions heavily depend on the materials flow stress at cutting areas such as high deformation rates and temperatures and variable friction characteristics at tool-chip interface which are not completely understood and need to be determined. Materials property at local shear band is very complex in the precision machining which makes it difficult to build up real robust flow stress model fitting for manufacturing process. Most of the energy consumption limited to local cutting area and transformed into heat which complicated the distribution of temperature at the local deformation area. The temperature plays an important role in the unstable chip flow. Larger plastic deformation rate and the intense friction at the tool-chip interface increase the heat generation rate and lead to the material softening thus decreasing the strain hardening ability and instability of materials flow. Therefore, the instability of shear behavior is directly induced by materials flow. Presently, researchers can't build up reasonable materials consititutive relationship which can characterize strain rate and the temperature and reflect the variation of materials property in the precision machining process.

Sound theoretical models based on atomic level material behavior are far from being accomplished. Semi-empirical constitutive models are widely utilized. Several material constitutive models are used in Finite Element (FE) simulation of metal cutting, including rigid-plastic, elasto-plastic, viscoplastic, elasto-viscoplastic and so on. These models take into account the high strains and temperatures reportedly found in metal cutting. Among others, the most widely used is the Johnson and Cook[7] (JC) model which is a thermo-elasto-visco-plastic material constitutive model expressed as follows:

$$\bar{\sigma} = \left[A + B(\bar{\varepsilon})^n \right] \left[1 + C \ln\left(\frac{\dot{\bar{\varepsilon}}}{\dot{\bar{\varepsilon}}_0} \right) \right] \left[1 - \left(\frac{T - T_0}{T_m - T_0} \right)^m \right] \tag{1}$$

here A is the initial yield stress of the material at the room temperature, strain rate 1/s and $\bar{\varepsilon}$ represents the equivalent plastic strain. The equivalent plastic strain rate $\dot{\bar{\varepsilon}}$ is normalized with a reference strain rate $\dot{\bar{\varepsilon}}_0$. Temperature term in JC model reduces the flow stress to zero at the melting temperature of the work materials, T_m, leaving the constitutive model with no temperature effect. In general, the parameters A, B, C, n, and m of the model are fitted to the data obtained by several material test conducted at low strains and strain rates and at room temperature as well as Split Hopkinson Pressure Bar (SHPB) test at strain rates up to 1000/s and at temperatures up to 600 °C. JC model provides good fit for strain-hardening behavior of metals and it is numerically robust and can easily be used in FE simulation models.

Besides, there are two major problems with the use of the discussed model and its method of the determination of its constants. First, only few laboratories and specialist in the world can conduct SHPB testing properly, assuring the condition of dynamic equilibrium. None of the known tests in metal cutting was carried out in these laboratories. Second, the high strain rate in metal cutting is rather a myth than reality. Third, the temperature in the so-called primarily deformation zone where the complete plastic deformation of the work materials takes place can hardly exceed 250 ºC. It is understood that the mechanical properties of the work material obtained at room temperature are not affected by this temperature so metal cutting is a cold working process, although the chip appearance can be cherry-red. Fourth, it is completely unclear how to correlate the properties of the work materials obtained in SHPB uniaxial impact testing with those in metal cutting with a strong degree of stress triaxiality.

3. The chip separation criterion on different materials used in the FEM

Presently, two FE methods exist for analyzing the precision machining process. In the first method, it is assumed that the chip formation is continuous and the shape of the chip is known in advance. Thus, the process is analyzed as a steady-state process. This method is called Eulerian method. In this method, a chip separation criterion is not required. In the second method, the process is analyzed from the beginning to the steady state chip formation. This is called Updated Lagrangian Formulation. In this method, a chip separation criterion is required to predict the chip geometry. Early applications of finite element method to the machining process were mainly Eulerian method. The main objective of many of these studies was to predict the temperature distribution and therefore, the determination of deformation and stress fields was only an intermediate step. These studies considered the machined material as rigid-plastic. But, later applications of Eulerian formulation to machining process also included viscoplastic effects. All of these applications have considered only orthogonal machining. The first finite element study of the machining process using an modified Lagrangian Formulation was made by Strenkowski and Carrol[8]. A critical value of the equivalent plastic strain was used to model the separation of a chip. Later on, several researchers used the Updated Lagrangian Formulation for analyzing two- and three-dimensional machining processes. The criterion used for chip separation has been based on controlled crack propagation or some geometrical considerations. Remeshing technique has been used to simulate the chip formation.

As the size of the material removed decreases in the precision machining, the probability of encountering a stress-reducing defect decreases. There are some new disciplines dominate the chip separation process. The metal cutting process is different from general metal forming process as there are always accompanied with chip separation or materials removal phenomenon. The separation of chip is of utmost important about numerical simulation of precision machining. The simulation results can only be meaningful only if the reasonable chip separation criteria which can reflect materials mechanical and physical property (such as morphology of chip, force, temperature and the residual stress etc.) were applied in the simulation model. Besides, the criterion for chip separation should be invariant for definite materials but not change with the different working conditions. In the metal cutting process, some kinds of materials may generate continuous chip while others may generate saw-like chip thus different materials fracture criteria should be included in the finite element model.

Presently, there are two kinds of chip separation criteria, namely, the geometric criterion and the physical criterion. Materials removal (chip separation) using geometric criterion is realized through the variation of size of deformable body. On the other hand, the physical criterion is based on if some key physical parameters approached the critical value, these physical criterion includes effective plastic strain criterion, strain energy density criterion and the fracture stress criterion and so on.

3.1 Fracture mechanics criterion

3.1.1 Stress intensity factor

In reality, chip separation process can be assumed as the formation and development of crack. Under what conditions and what manners can the materials be cut off is closely related with the fracture criterion[2]. Consider plane crack extending through the thickness of flat plane. There are three independent kinematic movements of the upper and lower crack surfaces with respect to each other. These three basic modes of deformation are illustrated in figure 1, which presents the displacements of the crack surface of a local element containing the crack front. Any deformation of the crack surface can be viewed as a superposition of these basic deformation modes, which are defined as follows:

1. Opening mode, the crack surfaces separate symmetrically with respect to the planes xy and xz
2. Sliding mode, the crack surfaces slide relative to each other symmetrically with respect to the planes xy and skew-symmetrically with respect to plane xz
3. Tearing mode, the crack surfaces slide relative to each other skew-symmetrically with respect to both planes xy and xz.

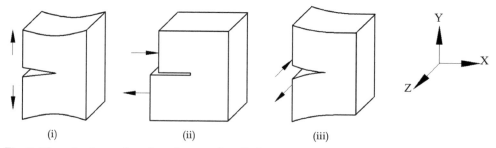

(i)   (ii)   (iii)

Fig. 1. Three basic modes of crack extension (i) Opening mode; (ii) Sliding mode; (iii) Tearing mode

The stress and deformation fields associated with each of these three deformation modes will be determined in the sequel for the case of plane strain and generalized plane stress. Solid materials is defined to be in a state of plane strain parallel to the plane xy if

$$u=u(x,y), \; v=v(x,y), \; w=0 \tag{2}$$

where u, v, w denote the displacement components along the axes x, y and z. Chip separation originated from crack while the static, stable or extension of the crack are all closely related with the distribution of stress field around the crack. The study of stress field near the crack tip is of great important as this field govern the fracture process that takes place at the crack tip.

a. Opening mode

Infinite plate with a crack of length 2a subjected to equal stresses σ at infinity is give by

$$Z_I(z) = \frac{\sigma z}{\sqrt{z^2 - a^2}} \tag{3}$$

If we place the origin of the coordinate system at the crack tip z=a through the transformation

$$\varsigma = z - a \tag{4}$$

Then the equation (3) takes the form

$$Z_I = \frac{\sigma(\varsigma + a)}{\sqrt{\varsigma(\varsigma + 2a)}} \tag{5}$$

using polar coordinates, r and θ we have

$$\varsigma = re^{i\theta} \tag{6}$$

the stress near the crack tip can be derived as follows:

$$\sigma_x = \frac{K_I}{\sqrt{2\pi r}}\cos\frac{\theta}{2}\left(1 - \sin\frac{\theta}{2}\sin\frac{3\theta}{2}\right) \tag{7}$$

$$\sigma_y = \frac{K_I}{\sqrt{2\pi r}}\cos\frac{\theta}{2}\left(1 + \sin\frac{\theta}{2}\sin\frac{3\theta}{2}\right) \tag{8}$$

$$\tau_{xy} = \frac{K_I}{\sqrt{2\pi r}}\cos\frac{\theta}{2}\sin\frac{\theta}{2}\cos\frac{3\theta}{2} \tag{9}$$

$$u = \frac{K_I}{4G}\sqrt{\frac{r}{2\pi}}\left[(2\beta - 1)\cos\frac{\theta}{2} - \cos\frac{3\theta}{2}\right] \tag{10}$$

$$v = \frac{K_I}{4G}\sqrt{\frac{r}{2\pi}}\left[(2\beta + 1)\sin\frac{\theta}{2} - \sin\frac{3\theta}{2}\right] \tag{11}$$

$$w = 0 \tag{12}$$

here σ_x, σ_y and τ_{xy} are the stress component, u, v and w are the displacement component, G is the shear modulus, μ is the poisson ratio, $\beta = 3 - 4\mu$. The K_I is the stress intensity factor and expresses the strength of the singular elastic stress field. As put forward by Irwin[9], equation (7) ~ (9) applies to all crack tip stress fields independently of crack/body geometry and the loading conditions. The stress intensity factor depends linearly on the applied load and is a function of crack length and the geometrical configuration of the cracked body.

$$K_I = \lim_{|\varsigma| \to 0} \sqrt{2\pi\varsigma}Z_I \tag{13}$$

Equation (13) can be used to determine the K_I stress intensity factor when the Z_I is known.

b. Sliding mode

Following the same procedure in the previous case, and recognizing the general applicability of the singular solution for all sliding mode crack problems, the following equations for stresses and displacements are obtained:

$$\sigma_x = -\frac{K_{II}}{\sqrt{2\pi r}} \sin\frac{\theta}{2}\left(2 + \cos\frac{\theta}{2}\cos\frac{3\theta}{2}\right) \tag{14}$$

$$\sigma_y = \frac{K_{II}}{\sqrt{2\pi r}} \sin\frac{\theta}{2}\cos\frac{\theta}{2}\cos\frac{3\theta}{2} \tag{15}$$

$$\tau_{xy} = \frac{K_{II}}{\sqrt{2\pi r}} \cos\frac{\theta}{2}\left(1 - \sin\frac{\theta}{2}\sin\frac{3\theta}{2}\right) \tag{16}$$

The K_{II} is the sliding mode stress intensity and can be obtained as following

$$K_{II} = \lim_{|\varsigma|\to 0} i\sqrt{2\pi\varsigma}Z_{II} \tag{17}$$

c. Tearing mode

$$K_{III} = \lim_{|\varsigma|\to 0} \sqrt{2\pi\varsigma}Z_{III} \tag{18}$$

The stress intensity factor is a fundamental quantity that governs the stress field near the crack tip. Several methods have been used for the determination of stress intensity factors as listed following:

a. Theoretical method (Westergaard semi-inverse method and method of complex potentials)
b. Numerical method (Green's function, weight functions, boundary collocation, alternating method, integral transforms, continuous dislocations and finite element method)
c. Experimental method (photoelasticity, holography, caustics)

Theoretical method is generally restricted to plates of infinite extent with simple geometrical configurations of cracks and boundary conditions. For more complicated situations one must result to numerical or experimental methods.

The stress intensity factor is one of the key parameters for characterizing stress field around crack, which can be used as the criterion for crack extension.

1. Single mode criterion

The single mode criterion can be expressed as follows:

$$K_I \geq K_{IC} \ , \ K_{II} \geq K_{IIC} \ , \ K_{III} \geq K_{IIIC} \tag{19}$$

here K_{IC}, K_{IIC}, K_{IIIC} are the fracture toughness of I, II and III modes separately, which is also the inherent property of materials.

2.   Mixed mode criterion

The mixed mode criterion can be acquired using Ellipsoid Criterion:

$$\left(\frac{K_I}{K_{IC}}\right)^2 + \left(\frac{K_{II}}{K_{IIC}}\right)^2 + \left(\frac{K_{III}}{K_{IIIC}}\right)^2 \geq 1 \tag{20}$$

3.1.2 *J*-integral theory

The stress intensity factor can only be applied to small yield around crack tip, other appropriate parameters should be developed to evaluated the large fracture strength. Rice[10] introduced path independent line integral as the elastic-plastic parameter for characterizing the status of crack which also named as *J*-integral. Hutchinson[11] and Rice and Rosengren[12] showed that *J* uniquely characterizes crack tip stress and strains in nonlinear materials. Thus the *J* integral can be viewed as both an energy parameter and a stress intensity parameter. After that, many researchers investigate the *J*-integral which establish the theoretical foundation of the path independent *J*-integral and its use as a fracture criterion. Presently, the main efforts in the study of elastic-plastic fracture mechanics is building up the evaluating method on fracture strength using *J*-integral while the yield materials around crack tip can be considered as non-linear elastic materials.

As for crack in the nonlinear elastic continuum medium, Rice[10] found that the integral around crack tip is path independent and is given by:

$$J = \int_{\Gamma}\left(wdy - T_i\frac{\partial u_i}{\partial x}ds \right) \tag{21}$$

here w is the strain energy density, T_i is the component of the traction vector, u_i is the displacement vector component and ds is a length increment along the contour Γ. The stress energy density is defined as:

$$w = \int_{0}^{\varepsilon_{ij}} \sigma_{ij}d\varepsilon_{ij} \tag{22}$$

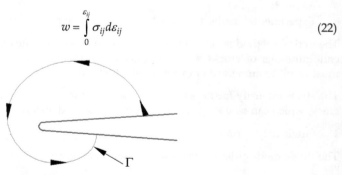

Fig. 2. Arbitrary contour around the tip of a crack

here σ_{ij} and ε_{ij} are the stress and strain tensors separately. The traction is a stress vector normal to the contour. That is, if we were to construct a free body diagram on the material inside of the contour, T_i would define the normal stress acting at the boundaries. The components of the traction vector are given by:

$$T_i = \sigma_{ij} n_j \tag{23}$$

here n_j is the component of the unit vector normal to Γ.

As for linear elastic materials, there some relationship as follows:

$$J = \frac{\kappa+1}{8\mu}\left(K_I^2 + K_{II}^2\right) + \frac{1}{2\mu}K_{III}^2 = G \tag{24}$$

As for nonlinear elastic materials, the system potential enclosed by curve Γ can be computed as follows:

$$\Pi = \int_\Gamma W(\varepsilon)dA_\Gamma - \int_\Gamma p_j u_j d\Gamma \tag{25}$$

Therefore

$$J = -\frac{\partial\Pi}{\partial a} \tag{26}$$

here a is the crack length. The J integral is essentially variation rate of system potential energy which is mainly transform into irreversible plastic work. If the work needed to extend crack a unit length is a constant, then the J integral based elastic-plastic fracture criterion can be deduced. It is because the J integral can be used to characterize the elastic plastic stress field solved by deformation theory that the J integral is selected as elastic plastic fracture criterion.

In 1968, Hutchinson[11], Rice and Rosengren[12] investigated the elastic plastic stress field around crack using deformation theory and acquired singular solution as follows:

$$\sigma_{ij} = \left(\frac{J}{a\varepsilon_Y \sigma_Y Ir}\right)^{\frac{1}{m+1}} \tilde{\sigma}_{ij} \tag{27}$$

$$\varepsilon_{ij} = a\varepsilon_Y \left(\frac{J}{a\varepsilon_Y \sigma_Y Ir}\right)^{\frac{m}{m+1}} \tilde{\varepsilon}_{ij}(\theta) \tag{28}$$

$$u_i = \left(\frac{J}{a\varepsilon_Y \sigma_Y Ir}\right)^{\frac{m}{m+1}} r^{\frac{m}{m+1}} \tilde{u}_i(\theta) \tag{29}$$

here I is definite integral of θ, $\tilde{u}_i$ is a function of θ. In reality, it is difficult to solve the J integral using equation (27) ~ (29) because of the complex regular expression of $\tilde{\sigma}_{ij}$, $\tilde{\varepsilon}_{ij}$ and $\tilde{u}_i$. The numerical method and the energy method are the two practical solutions. The numerical method mainly makes use of elastic-plastic finite element method and integrates along several paths around crack tip and acquires the J integral. The final J integral can be computed as follows:

$$J = \sum \frac{J_i}{n} \tag{30}$$

here J_i is the J integral corresponding to path Γ_i, n is the number of integrate path. The integrate path is generally continuous smooth curve which can reduce the error resulted by the discontinuous surface force.

3.2 Geometrical criterion

The geometrical criterion mainly takes effect through judging if the geometrical size of materials exceeding the criterion. Figure 3 shows the geometrical model in which a separation line is defined. The nodes at the chip side and the nodes at workpiece side are overlapped at the beginning. But the separation of two nodes occurs when the distance D between the tool cutting edge (point d, in Figure 3) and the node immediately ahead (node a) becomes less than a predefined critical value thus the machined surface and the chip bottom are generated.

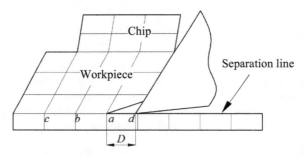

Fig. 3. Geometrical criterion model

Usui and Shirakashi[13] first put forward the geometrical criterion and found it is a stable criterion. Komvopoulos and Erpenbeck[14] pointed that there should be enough distance between tool tip and the overlap point to prevent the convergence problem resulted by the excessive distortion of finite element mesh. Zhang and Bagchi[15] brought forward that the geometrical distance should be less than 30 percent to 50 percent of element length. Furthermore, they also put up a new geometrical separation criterion which is based upon the ratio of geometrical distance to depth of cut which is equivalent to the microscopic fracture mechanics criterion.

The geometrical criterion is simple to be used in the FE computation. However, the distance (D) between tool tip and the separation point is closed to zero which result in the difference between the set value of D with the reality. The selection value of D will have a great influence upon the convergence of FE simulation and only the experienced researcher can deduce appropriate valuable critical value. In addition, the separation line which separates the mesh of chip and that of the workpiece should be built up in advance. Figure 4 shows the FE simulation of precision machining process based on geometrical separation criterion.

4. Materials deformation behavior in the precision machining

The depth of cut in the precision machining is very small, chips are formed at very narrow regions. The work material is subjected to extremely high plastic deformation and the strain rates can reach the values of about 10^5 s^{-1}. The large strain and high strain rate plastic deformation evolves out of hydrostatic pressure that travels ahead the tool as it pass over. The zone has, like all plastic deformations an elastic compression region that becomes the plastic compression region as the field boundary is crossed. The plastic compression generates dense dislocation tangles and networks which lead to the materials shear after the materials experience fully work hardened. The theory of micro-plasticity, which mathematically describes the stress and strain at small scale, is adopted to calculate the distributions of stress and strain in the distorted bodies.

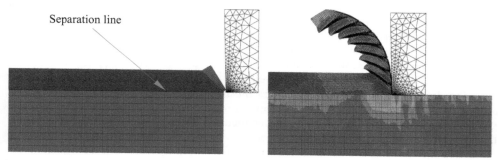

Separation line

Fig. 4. FE simulation based on geometrical separation criterion

4.1 Plastic deformation and chip formation in the precision machining titanium alloy

The numerical analysis method applied to materials cutting process can be divided into two categories, namely, the elastic-plastic FEM and the rigid-plastic FEM. Furthermore, thermo-elastic FEM and the thermo-rigid FEM are introduced if the temperature and the velocity are considered in the materials processing technology. The simulation results are almost same whether the problem analysed by either elastic-plastic FEM or rigid-plastic FEM if the size of the workpiece and the amount of discreted element are same for these two methods. The elastic-plastic FEM mainly applied to solve the residual stress and the elastic recovery while the rigid-plastic FEM cannot solve this type of problems as it ignored elastic deformation and thus it has higher solution efficiency.

In this research work, the commercial finite element analysis package (Advantedge®) is utilized to gain good understanding of the materials deformation behavior underlying machining of titanium alloy. Among the different alloys of titanium, Ti-6Al-4V is by far the most popular with its widespread use in the chemical, surgical, ship building and aerospace industry. The primary reason for wide applications of this titanium alloy is due to its high strength-to-weight ratio that can be maintained at elevated temperatures and excellent corrosion and fracture resistance. On the other hand, Ti-6Al-4V is notorious for poor machinability due to its low thermal conductivity that causes high temperature on the tool face, strong chemical affinity with most tool materials, which leads to premature tool failure, and inhomogeneous deformation by catastrophic shear that makes the cutting force fluctuate and causes tool wear, thereby aggravating tool-wear and chatter. This poor machinability has limited cutting speed to less than 60 m/min in industrial practice. Numerical analysis of Ti-6Al-4V machining process using finite element method is of great importance on understanding the physical essence and optimizing the machining technique parameters.

4.2 Finite element formulation

The FEM mesh is constituted by elements that cover exactly the whole of the region of the body under analysis[3]. These elements are attached to the body and thus they follow its deformation. Metal cutting process is a large deformation and finite strain related elastic-plastic process. Therefore, both nonlinear material property and the nonlinear geometry property ought to be considered in the numerical analysis. Presently, typical finite element formulations used in metal cutting include Lagrangian or Eulerian method. Lagrangian formulation bases upon the original geometry which also termed as particle coordinates description, Eulerian formulation bases upon the deformed geometry which termed as floating coordinate description. These formulations are particularly convenient when unconstrained flow of material is involved, i.e., when its boundaries are in frequent mutation. In this case, the FE mesh covers the real contour of the body with sufficient accuracy. On the other hand, the Eulerian formulation is more suitable for fluid-flow problems involving a control volume. In this method, the mesh is constituted of elements that are fixed in the space and cover the control volume. The variables under analysis are calculated at fixed spatial location as the material flows through the mesh. This formulation is more suitable for applications where the boundaries of the region of the body under analysis are known a prior, such as in metal forming.

Although both of these formulations have been used in modelling metal cutting processes, the Lagrangian formulation is more attractive due to the ever-mutating of the model used. The Eulerian formulation can only be used to simulate steady state cutting. As a result, when the Lagrangian formulation is used, the chip is formed with thickness and shape determined by the cutting conditions. However, when one uses the Eulerian formulation, an initial assumption about the shaped of the chip is needed. This initial chip shape is used for a matter of convenience, because it considerably facilitates the calculations in an incipient stage, where frequent problems of divergence of algorithm are found.

The Lagrangian formulation, however, also has shortcomings. First, as metal cutting involves severe plastic deformation of the layer being removed, the elements are extremely distorted so the mesh regeneration is needed. Second, the node separation is not well defined, particularly when chamfered and/or negative rake or heavy-radiused cutting edge tools are involved in the simulation. Although the severity of these problems can be reduced to a certain extent by a denser mesh and by frequent re-meshing, frequent mesh regeneration causes other problems.

These problems do not exist in the Eulerian formulation as the mesh is spatially fixed. This eliminates the problems associated to high distortion of the elements, and consequently no re-meshing is required. The mesh density is determined by the expected gradients of stress and strain. Therefore, the Eulerian formulation is more computationally efficient and suitable for modelling the zone around the tool cutting edge, particularly for ductile work materials. The major drawback of this formulation, however, is that the chip thickness should be assumed and kept constant during the analysis, as well as the tool–chip contact length and contact conditions at the tool–chip and tool–workpiece interfaces. As the chip thickness is the major outcome of the cutting process that defines all other parameters of this process so it cannot be assumed physically. Consequently, the Eulerian formulation does not correspond to the real deformation process developed during a real metal cutting process.

The Lagrangian formulation[16] under finite deformation is as follows:

$$\{p_t^\alpha\} = \iiint_{V_0} \left[B'_{ij} \right]^T S_{ij} dV_0 + \iiint_{V_0} \left[B''_{ij} \right]^T S_{ij} dV \tag{31}$$

where $\{p^\alpha\}$ denotes the column vector of external force exerted at the discrete element nodes, $[B']$ is the geometry matrix in the case of finite strain conditions and the $[B'']$ is the additional item induced by the geometric nonlinear conditions.

4.3 Finite element model and simulation results

The corresponding mesh is refined in some region as severe plastic deformation may be induced under material surface which is shown in figure 5. The most fundamental and crucial characteristic of metal cutting process lies in the formation of chip. In reality, the chip is not exactly "cut" but "sheared" away from the work material which forms a clear distinction between machining plastic metal and other materials. Figure 6 shows the chip formation process during precision machining of titanium alloy. Chip formed with the tool approaching the material from the right side and the chip flow in curved fashion. When the original chip thickness or feed rate or depth of cut is compared with the chip thickness after cutting, the deformation can be clearly observed. This deformation is fundamental for the

Material	Titanium
Size (mm)	$0.4 \times 0.01 \times 0.1$
Physical Property	Elastic-Plastic Solid
Depth of Cut(μm)	5
Speed(mm/s)	200
Temperature($^\circ C$)	20

Table 1. FEM simulation parameters

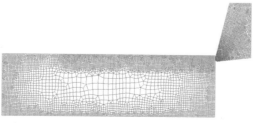

Fig. 5. FE simulation model

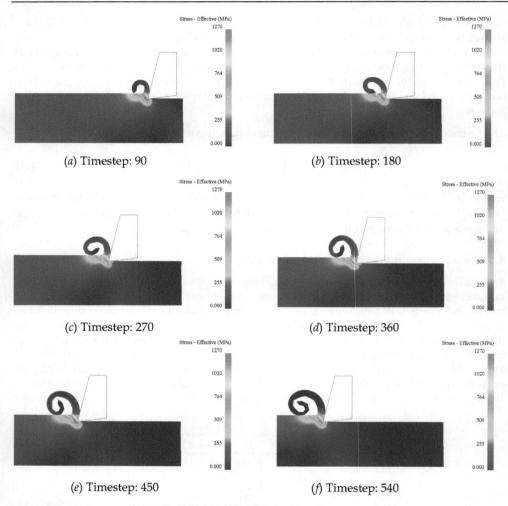

(a) Timestep: 90   (b) Timestep: 180

(c) Timestep: 270   (d) Timestep: 360

(e) Timestep: 450   (f) Timestep: 540

Fig. 6. FE simulation of precision machining of titanium alloy

metal cutting process and involves large deformations of materials with very large strains and very high strain rates. The produced chip is in contact with the tool face in a highly pressurized zone causing sticking friction which transforms to sliding friction further up on the tool face. A large amount of heat is generated in the cutting zone as a result of plastic work and friction causing temperature rise in the tool and chip.

There are three main plastic deformation areas in this precision machining process as shown in figure 6, namely, the first plastic deformation region, which dominates the kind and the morphology of the chip and generated large amount of heat, the degree of plastic deformation is closely related with materials stress-strain relationship; the second plastic deformation region where the intense tribology process is generated between bottom of chip and rake face of cutting tool; the third plastic deformation region where the tribology behavior is generated between materials machined surface and the clear face of cutting tool. With the cutting in of tool, the elastic deformation is initially induced at the contact interface between cutting tool and materials. After that the titanium alloy becomes going into yield state with the further successively feeding of cutting tool and the plastic deformation region gradually comes into being ahead of cutting tool. The successive feeding of cutting tool results in the contraction of the elastic deformation and expansion of plastic deformation. The full contact between cutting tool and workpiece comes into being and the elastic-plastic deformation is generated. The simulation results show that fairly concentrated shear separates the nearly unstrained work materials from the fully strained chip. But no obvious region of secondary deformation is generated close to the rake face of tool. The contact length between rake face of cutting tool and the bottom is very small which also justifies most of the cutting process are accomplished by the local tool tip.

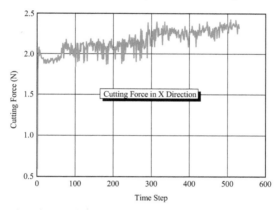

Fig. 7. Simulation results of cutting force

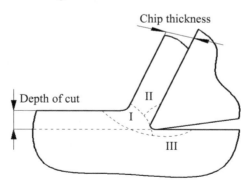

Fig. 8. Deformation area in the metal cutting

Metal cutting process at nanometer scale involves plastic deformation in small localized regions where opposing surface contact or in the interior of workpiece materials. As for chip formation, the single-shear plane model and practically all its "basic mechanics" have been known since nineteenth century and referred as the Merchant (or Ernst-Merchant model) model[1]. This model has been the basis for most of the present metal cutting analysis. The first orthogonal model was brought forward in 1937 by Piispanen[1] and termed as card model. In this model, the material cut is assumed as a deck of cards inclined to the cutting direction which is shown in figure 9. Merchant assumed the chip to be formed over an infinite thin plane called shear plane. This shear plane starts from the cutting edge of the tool and crosses the chip on an angle with the cutting direction, which is termed as shear angle. When the chip passes the shear plane it is sheared away from the workpiece and increases in thickness. In this simulation, no single shear plane is observed in the whole precision machining process. On the other case, there some maximum stress band is continuously generated in front of cutting tool. This shear band possesses irregular geometry shape which extends from first deformation region to third deformation region.

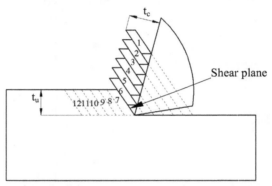

Fig. 9. Card of cutting process

A zone of plastic deformation extends underneath the machined surface. This subsurface deformation will result in compressive stresses in the machined surface. Though the stress patterns are those with the load applied by the tool still present, elastic recovery caused by the unloading of the tool is not expected to significantly change the stress distribution close to the free surface. So the stress in the machined surface sufficiently far away from the tool can be taken to be the residual stress. The location of the nodes along the machined surface when compared with the location of tool cutting edge yields information about the elastic recovery of the machined surface after it passes under the tool. The elastic spring-back of the machined surface is found to be far less than the radius curvature of cutting edge which justify that most of the material in front of the rounded cutting edge is actually pushed ahead of the tool and not into the machined surface.

The simulation results also shows that the continuous internal curling chip is generated under current working conditions. At the beginning, part of chip adjacent to the tool tip begins to curl and form helix circle with small radius. After that, the larger helix circle surround the previous small one is gradually formed with the feeding of the cutting tool. The deformation coefficients ($\xi = \dfrac{t_c}{t_u}$) is gradually increased in this process which result in the increasing of cutting force (figure 7). The stress along the free surface (back) of chip is tensile. It is also tensile along the surface of chip which has moved out of the contact with the tool rake face (front) while the σ_{yy} in the middle of the chip is compressive. Such a distribution of stress is the critical factor to develop initial formation of chip.

Presently, the hypotheses propounded by various researchers to explain the curvature of the chip include (i) The cutting moment causes the chip to bend; (ii) The 'crushing' of chip in the secondary shear zone and the resultant acceleration of the work material in moving through the secondary shear zone causes the chip to lengthen along this side (the front side). This can also results in a curvature of the chip which is similar to the curvature of a bimetallic strip; (iii) The shear plane is curved in such a way that the shear plane angle is smaller near the exit of shear plane. Thus the chip velocity on the back side is smaller than the average chip velocity which causes the chip to curl.

The bending moment on the chip considered as a beam would result in compressive stress along the free surface (back) of the chip if hypothesis (i) was true. Crushing of the chip in the secondary shear zone will result in compressive σ_{yy} in the front (underside) of the chip. Only a curved shear plane would result in a stress distribution similar to that given by the finite element analysis, while simultaneously accounting for curl of the chip. It should be noted that though the chip does accelerate (due to secondary shear) as it flows along the rake face of tool, this is just an accessory to chip curl and not the cause of chip curl. The reason for the curvature of the shear plane can be found from a detailed analysis of the stress distribution in the zone of plastic deformation. Work in this direction is in progress.

5. Conclusion

With the increasing of high quality and accuracy of modern automated machining technology, numerical simulation of machining technology such as FEM is starting to emerge. The FEM based virtual machining simulation has the capability of calculating the results of process variables about the precision machining process used for optimization the cutting process thus providing many benefits to the metal cutting application. Presently, FEM is mainly of use to mechanical and materials engineering, as a tool to support process understanding, materials machinability development and tool design. The research efforts show that the model used in FEM of precision metal cutting process should be adequate to the process. But the concept of FE model should be broadened in order to embrace important facets physics including uncertainty, which has been axiomatized out of modern cutting research. Breakthrough in these directions will have considerable impact by making metal cutting simulation useful for practical optimization of various metalworking operations including the cutting and machine tools, the metal working fluids and fixtures and so on.

6. References

[1] M.C. Shaw, Metal Cutting Principles, Oxford Science, Oxford, 1984
[2] T.L. Anderson, Fracture Mechanics – Fundamentals and Applications, CRC Press, New York, 1995
[3] J. Chaskalovic, Finite Element Methods for Engineering Sciences, Springer-Verlag, 2008

[4] E. Ceretti, P. Fallböhmer, T. Altan, Application of 2D FEM to chip formation in orthogonal cutting. J Mater Process Technol, 1996, 59(1~2): 160~180

[5] Shih Albert, J.M. Chandrasekar, S. Yang, T.Y. Henry, Finite element simulation of metal cutting process with strain-rate and temperature effects, ASME Prod Eng Div Publ PED, 1990, 43: 11~24

[6] Xuda Qin, Pingyu Zhu, Jiangang Zhang etc., Modeling and Analysis on Surface Temperature of Al Alloy in Plunge Milling, China Mechanical Engineering, 2007, 18(17): 2041-2042

[7] McClain, B., Thean, W., Maldonado, G.I. etc., Finite element analysis of chip formation in grooved tool metal cutting, Mach Sci Technol, 2000, 4(2):305~316

[8] G. R. Johnson, W. H. Cook, A constitutive model and data for metals subjected to large strain, high strain rates and high temperature, 1983, Proceedings of seventh international symposium on ballistic, The Hague, Netherlands

[9] J. S. Strenkowski, J. T. Carroll, A finite element model of orthogonal metal cutting, ASME Journal of Engineering for Industry, 1985, 107: 349~354

[10] G. R. Irwin, Fracture, Encyclopedia of Physics, VI (Elasticity and plasticity), Springer-Verlag, 551~590

[11] J. R. Rice, A path independent integral and the approximate analysis of strain concentration by notches and cracks, Journal of Applied Physics, 1968, 35: 379~386

[12] J.W. Hutchinson, Singular behavior at the end of a tensile crack tip in a hardening material, Journal of the Mechanics and Physics of Solids, 1968, 16: 13~31

[13] J. R. Rice, G. F. Rosengren, Plane strain deformation near a crack tip in a power law hardening material, Journal of the Mechanics and Physics of Solids, 1968, 16: 1~12

[14] E. Usui, T. Shirakashi, Mechanics of machining – From descriptive to predictive theory, On the art of cutting metals – 75 years later a tribute to F W Taylor, ASME PED-7 1982, 13~30

[15] F. K. Komvopoulos, S. A. Erpenbeck, Modelling of orthogonal metal cutting, Trans of ASME J. Eng. Ind, 1991, 113: 253~267

[16] B. Zhang, A. Bagchi, Finite element simulation of chip formation and comparison with machining experiment, Computational method in material processing, ASME publication, PED, 1992, 61: 61~74

[17] S. S. Xie, Z. T. Wang, Finite Element Numerical Simulation of Metal plastic Deformation, Metallurgical Industry Press, Beijing, 1997

7

Energy Dissipation Criteria for Surface Contact Damage Evaluation

Yong X. Gan

Department of Mechanical, Industrial and Manufacturing Engineering,
College of Engineering, University of Toledo,
USA

1. Introduction

This chapter presents the energy dissipation approach for analyzing surface contact damages in various materials, including composite materials. As known, surface contact is a very common phenomenon, which can be found in daily life and many scientific and engineering problems. The contact of different bodies can be modeled as indentation. Analysis of indentation and modeling of the deformation states of indented materials are often difficult because of the complexity of stress distributions within indentation zones. It is also very difficult to evaluate stress states in regions underneath an indented zone. Instrumental indentation has been performed on various materials including composite materials. Experimental studies on indention of coatings and brittle materials have been reported extensively, but the criterion for evaluating the extent of damage is not unified. Ductile materials deform relatively stable in indentation processes. While brittle materials are sensitive to compressive contact loadings in view of the formation of surface cracks. Therefore, it is difficult to find a unified stress or strain based damage criterion to characterize the damage evolution. Energy dissipation analysis may be more accurate to describe the deformation behavior of such materials. Specifically, under wedge indentation, the analysis should be investigated because the stress field has the singularity which limits the applicability of the strength criterion. In this chapter, the load-displacement relations with elastic-plastic responses of the materials associated with the indentation processes will be obtained to calculate the hysteresis energy. Lattice rotation measurement using electron backscatter diffraction (EBSD) technique will be performed in the region ahead of the indenter tip to measure the dimension of the contact damage zone (CDZ) and the results will be used to define the length scales in contact deformation. A unified criterion using the hysteresis energy normalized by the length scales will be established.

Damage evolution in composite materials is very sensitive to the interaction of reinforcements and matrices in interface regions. For example, the development of damage in glass particle and fiber reinforced epoxy composite materials is strongly influenced by the interface debonding conditions [1]. However, the exact effect of bonding conditions on the performance of particle filled composite materials is still not fully understood. Kawaguchi and Pearson [2] reported that strong matrix-particle adhesion may lower the fatigue crack propagation resistance. While the studies on Si_3N_4 nanoparticle filled epoxy composites under sliding wear conditions showed that the strong interfacial adhesion between Si_3N_4 nanoparticles and the matrix reduced the wear rate of the composites [3]. Damage in the form of debonding in coated fiber reinforced composites under tension-tension cyclic load was investigated [4]. The bi-interfacial debonding (fiber/coating and coating/matrix) behavior was analyzed using a double shear-lag model. Based on this model, the debond growth rate and strain energy were calculated by finite element method. Non-uniform damage of coating materials was accounted in the analysis. There exists two-interface coupling in debonding. It was found that the strength and thickness of coating materials are the major factors controlling the bi-interfacial crack growth. Numerical simulation of progressive damage evolution in fiber reinforced composites was performed to understand interface stress statistics and the fiber debonding paths development [5]. A meso cell including several hundred inclusions was used to account for the micro structure statistics of the composites. Both the local stress and effective elastic moduli of disordered fibrous composites were computed.

Micromechanics based approaches have been used for debonding damage analysis [6-10]. Cavallini, Bartolomeo, and Iacoviello [6] investigated the damage in three different ferritic-pearlitic ductile cast irons with the main focus on graphite nodules debonding. Chan, Lee and Nicolella et al. [7] studied the near-tip fracture processes of nanocomposites under cyclic loads. It is found that particle bridging, debonding at the poles of particle/matrix interface, and crack deflection around the particles are the major micromechanics responses to cyclic loadings. Environmental conditions on the subcritical debond-growth rates were also examined [8]. Temperature and relative humidity are sensitive factors. Long term exposure to a moist environment resulted in the time-dependent decrease in adhesion between matrices and reinforcements. Three different interfacial damage models including the shear lag model, the linear degradation model and the modified power degradation model were used to describe the bond decay at steel/concrete interface [9]. The role of internal friction in resisting interfacial debonding was addressed. Micro-level damage in discontinuous fiber reinforced composites were found in the forms of fiber/matrix interfacial debonding and fiber failure [10]. The Weibull damage law was used to predict the microscopic damage behavior of composites with different fiber contents and orientations.

Crack initiation or small crack growth plays a critical role in interface debonding [11]. In small crack growth, plasticity-induced crack closure was observed, but the effect of crack closure in fatigue crack growth predictions was less than the estimation by the classical approaches [12]. In addition to crack closure, the shear deformation of matrix ahead of a small crack slows down the interfacial debonding rate [13]. Interface debonding controlled small crack growth behavior depends on the stress levels [14], and loading rate [15-16]. Microdebonding or subcritical debonding behavior is also dependent on surface chemistry [17] and temperature [18, 19]. To evaluate the surface chemistry effect, subcritical debonding of thin polymer layers from inorganic dielectrics was studied using selected amino- and vinyl- functional silane adhesion promoters [17]. Due to the surface modification, the failure occurs not at the interface but in a region very close to the interface. The effect of temperature on debonding is especially significant in metal matrix composite materials [20-34]. At elevated temperatures, thermomechanical fatigue accounts for the failure of these materials. Alternating plastic shearing of the interface takes place under combined mechanical and thermal stresses [18]. At low temperatures, metal matrices such as Al typically shows an initial hardening process, while at high temperatures, only cyclic softening is found [19].

Fatigue tests on reinforced titanium composites revealed various interface damage mechanisms [20-27]. Shear frictional sliding [20], interfacial debonding [21], fiber bridging [22], surface embrittlement [23], matrix ligament premature ductile shear [24], and crack deflection [25] are typical damage mechanisms observed. These damage mechanisms could occur simultaneously depending on loading modes, but debonding always exists and is considered as the major mechanism. A stress-based criterion for predicting the debonding behavior was proposed [22]. Rios, Rodopoulos and Yates [26] assessed the initial and final damage states caused by interface debonding and fiber bridging to determine the damage accumulation rates in SiC fiber reinforced titanium composite. Their method was used for damage tolerant fatigue design. Residual stiffness and the post-fatigued tensile strength as a function of microstructural damage were obtained through computer simulation, and the interfacial frictional stress and the critical crack length were also calculated [27]. Under combined thermal and mechanical fatigue loading, carbon fiber/Al and SiC fiber/Al composites were found to fail by a ratchetting mechanism, which is characterized by the progressive plastic deformation increasing with the number of cycles, even at stress levels far below the yield stress [28]. It is further found that the main phenomenon leading to composite failure is ratchetting at high load levels and interface degradation at low load levels.

Short crack growth behavior in steels containing different particle inclusions including Al_2O_3, MnS and Ti_3N_4 was studied by finite element method [29]. Crack-tip displacements and energy release rates were taken as the driving forces. It was found that the energy release rate is the highest for the Al_2O_3 inclusion case with a short through thickness crack. Li and Ellyin [30] studied the fatigue damage and the localization in Al_2O_3 particulate reinforced aluminum composites. The primarily damage forms are particle debonding, fractured particles and matrix cracks. Mesoscale reinforcement defects, such as a clump of large particles were also found causing damage localization. These defects were assumed to be the reason for short crack initiation and extension. In Murtaza and Akid's work on steel [31], it is reported that debonding at the matrix/inclusion interface is the major mechanism for the formation of short cracks. Stress redistribution at interfaces in alumina/aluminum multilayered composites was investigated [32]. The effects of interfacial debonding or of plastic slip in the metal phase adjacent to strongly bonded interfaces were considered. The results of stress measured around the crack reveal that debonding is much more effective than slip in reducing the stress ahead of the crack. Interaction of short fatigue crack with different types of particles was studied. Stronger interaction of fatigue crack with Si particles, as compared to SiC particles, was observed in particle reinforced A356 casting alloy [33].

Modeling fatigue debonding have been performed by many researchers [34-39]. In Gradin and BÄacklund's work [34], a unit cell model containing a steel bar and a co-centric epoxy cylinder was used to study the progressive de-bonding between the fiber and the matrix. Energy release rate was correlated to the interfacial debonding length. While in the work shown in [35-37], void formation and growth due to fatigue loading was characterized by the tensile stress at the interface. Three distinguishable debonding stages, two transient ones separated by a steady stage, were defined by Botsis and Zhao [38]. Stress intensity factor may be used to distinguish the steady and the transient stages because the total stress intensity factor was found to be approximately constant at the steady state. Debonding under different loading modes including mode I, mode II and mixed mode (I & II) was studied by Dessureautt and Spelt [39]. It was observed that the debonding rate was the greatest under mixed-mode conditions.

In this chapter, the emphasis on mechanics analysis will be put on the damage initiation and propagation from the debonding of particle/matrix interface. Both macro- and micro-scale analysis will be performed. The macroscale approach based on continuum mechanics will be used to obtain the stress field in the elastic-plastic region within the matrix in front of the debonded particle. Treating the debonded region as a crack, stress intensity solutions can be obtained. In the plastic zone just ahead of the debonded particle, the microscale approach will be used to find the stress solutions. In the classical plasticity theory, the material property at the crack tip is considered to be isotropic and the maximum stress in the plastic zone is assumed to be the yield strength of the material. In this work, the particle-matrix interaction is modeled as surface contact and multiscale approaches are used in the modeling and experiments.

Why the particle-matrix interaction and the debonding in the interface region can be modeled as contact damage under indentation load? The rationale is evident by examining the damage zone. As shown in Figure 1, the particle inclusion is debonded from pearlitic steel matrix. Along the main crack propagation direction (marked as x-direction), two distinct slip regions are found. These regions are denoted as *Region I* and *Region II*. In each of these regions, persistent slip lines are found. Although there are also some other slip zones around the particle, the predominant slip activities that determine the main crack speed are from *Region I* and *Region II*. Therefore, with a simplified model, the slip in these two regions can be seen as generated by indentation. The hard particle is equivalent to an indenter.

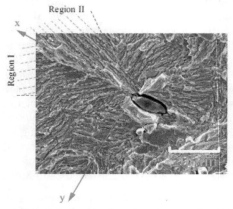

Fig. 1. Scanning electron microscopic image showing contact damage induced slip zones around a debonded particle. The main crack propagation is along x-axis.

2. Surface contact damage model

The first part of the modeling work is on the surface contact damage initiation using a microscale approach. Since the deformation state at the contact point is highly anisotropic, the deformation mechanism of single crystal plasticity is enforced in this stage. The deformation of the material in the indenter tip region due to the motion of dislocation on different slip systems will be described. Based on such a consideration, we assume that the stresses at the boundary between the elastoplastic region and the plastic zone propagate into the plastic zone. The magnitudes of the stress components are determined. The primary slip lines are assumed to be collinear with the dislocation motion directions. The second part of this section is specifically on the contact damage propagation. Once a short crack from the interface debonding starts growing, how to characterize the fatigue crack growth resistance becomes an important issue. A simulated crack (indenter penetration depth) is used to study the contact damage propagation kinetics. The specific energy of damage, a parameter which is used to characterize the resistance of the material to contact damage, is defined. The relationship between energy release rate and the specific energy of damage is established.

2.1 Contact damage initiation stage: microscale approach

It is assumed that the matrix is elastic-plastic so that in-plane slip is the prevailing plastic deformation mechanism. S is the unit vector parallel to the slip direction. N is the unit vector along the slip plan normal. To use indentation to simulate the debonding, the partial debonding and the fully debonded states, as shown in Figures 2(a) and 2(b), respectively, can be treated by the cases with indenter partially penetration and retreating. In order to find the stress solution, the debonded region is considered as a crack. A stress intensity approach is applied to find an approximate solution. Figures 2(c) shows both the global and the local coordinates for deriving the stress solutions in the slip regions.

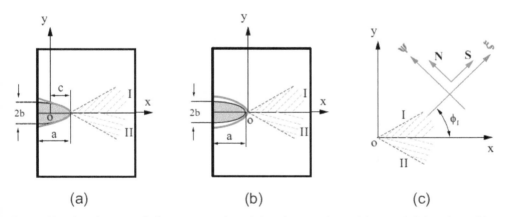

(a)   (b)   (c)

Fig. 2. Sketches for particle/matrix interface debonding analysis: (a) partial debonding, (b) complete debonding, (c) configuration of global, local coordinates related to the slip direction and slip plane normal vectors.

In a contact cycle, supposing that plane-strain conditions hold, the non-zero components of the stress field ahead of the particle are calculated by fracture mechanics as

$$\sigma_{xx} = \frac{K_I}{\sqrt{2\pi r}} \cos\left(\frac{\theta}{2}\right)\left[1 - \sin\left(\frac{\theta}{2}\right)\sin\left(\frac{3\theta}{2}\right)\right] \tag{1a}$$

$$\tau_{xz} = \tau_{zx} = \frac{K_I}{\sqrt{2\pi r}}\cos\left(\frac{\theta}{2}\right)\sin\left(\frac{\theta}{2}\right)\sin\left(\frac{3\theta}{2}\right) \tag{1b}$$

$$\sigma_{zz} = \frac{K_I}{\sqrt{2\pi r}}\cos\left(\frac{\theta}{2}\right)\left[1+\sin\left(\frac{\theta}{2}\right)\sin\left(\frac{3\theta}{2}\right)\right] \tag{1c}$$

$$\sigma_{yy} = \mu\left(\sigma_{xx} + \sigma_{zz}\right) \tag{1d}$$

where K_I is the stress intensity factor related to the particle shape.

Other stress components are zeros, i.e. $\tau_{xy} = \tau_{yx} = \tau_{yz} = \tau_{zy} = 0$. Assuming the material near the tip is fully plastic, the following yielding criterion holds

$$N\Sigma S = \pm\tau_i \tag{2}$$

where τ_i is the shear strength of the ith slip system, $\tau_i = \tau_I$ for *Region I* and $\tau_i = \tau_{II}$ for *Region II*. N is the surface normal of the slip plane, S is a unit vector along the slip direction. If the dislocation motion is along positive S, the right hand side takes positive τ_i, while in the case that the slip occurs along negative S, the negative sign is kept on the right hand side. Σ is the stress tensor. The components of N are N_x, N_y and N_z, and S has the components: S_x, S_y and S_z. Since only the in-plane slip is considered in this work, the z-components for both N and S are zeros. Therefore, the yield condition is

$$N_x\sigma_{xx}S_x + N_y\sigma_{yy}S_y = \pm\tau_i \tag{3}$$

where $S_x = \cos(\varphi_i)$, $S_y = \sin(\varphi_i)$, $N_x = -\sin(\varphi_i)$, $N_y = \cos(\varphi_i)$, $\varphi_i = \varphi_I$ for *Region I* and $\varphi_i = \varphi_{II}$ for *Region II*. Substituting these relations into Eq. (3) yields

$$-\frac{\sigma_{xx} - \sigma_{yy}}{2}\sin(2\varphi_i) = \pm\tau_i \tag{4}$$

Eq. (4) provides the yield function related to the slip angle and the stress field when the material is in a fully-plastic state. For the partial debonding case, along the radial line $\theta = 0$, the in-plane stresses are

$$\sigma_{xx} = \frac{K_I}{\sqrt{2\pi r^*}} \tag{5a}$$

$$\tau_{xy} = 0 \tag{5b}$$

$$\sigma_{yy} = \pm\frac{2\tau_i}{\sin(2\varphi_i)} + \sigma_{xx} \tag{5c}$$

where r^* is the distance from the origin to an arbitrary point on the $\theta = 0$ radial line, and $K_I = \frac{1.12}{\pi}\sqrt{\pi(a-c)\sigma_\infty}$.

Once the stress field along the radial line $\theta = 0$ is obtained, it is straightforward to find the stress state within the slip region. One of the ways is to follow the slip line analysis [40] to solve the stress components in *Region I* and *Region II*.

2.2 Contact damage propagation stage: macroscale approach

In this part, evaluation of the contact damage propagation behavior based on experimentally determined irreversible work and energy dissipation is presented. The energy dissipated into damage formation is considered as the indentation penetration driving force. A materials parameter, the specific energy of damage is used as the contact damage tolerance criterion as previously introduced for some materials in [41-46]. Correlation between the contact damage tolerance and the microstructure of the material is made.

Considering indenter penetration region and its surrounding damage zone in the material as a thermodynamic entity, the following relationship can be obtained based on entropy and energy balance considerations.

$$T\dot{S} = \left(J^* - a\gamma \right)\frac{da}{dN} + D \tag{6}$$

where T is the ambient temperature and $\dot{S}$ is the rate of change of the entropy of the system comprising the indenter penetration region and the surrounding damage zone. J^* is the energy release rate. γ is the specific damage of energy. a is the nominal indenter penetration depth or developed contact length between the indenter and the indented material. da/dN is the indenter penetration speed. N is the number of indentation cycles. D is the rate of energy dissipation into contact damage formation associated with the damage zone evolution.

At minimum entropy, $T\dot{S} = 0$. Eq. (6) can be rearranged as

$$\frac{da}{dN} = \frac{D}{a\gamma - J^*} \tag{7}$$

Under force control indentation conditions, the energy release rate J^* can be evaluated by

$$\frac{da}{dN} = \frac{1}{B}\frac{\partial P}{\partial a} \tag{8}$$

where P is the potential energy (area above the unloading curve) at the indenter penetration depth a, and B is the specimen thickness. The cyclic rate of energy dissipation, D associated with contact damage zone evolution can be evaluated by the difference between the hysteresis energy related to indentation and the hysteresis energy dissipated into the bulk of the material. It can be expressed as:

$$D = \frac{H_n}{B} \tag{9}$$

where H_n is the hysteresis energy. Rearranging Eq. (7) yields

$$\frac{J^{*}}{a} = \gamma - \frac{D}{a\left(\dfrac{da}{dN}\right)} \tag{10}$$

The quantities J^{*}, da/dN, and a, can be obtained from indentation experiments. The relationship expressed in Eq. (10) can be plotted in a two dimensional domain, directly giving the value of the specific energy of damage, γ, which is the intercept of the straight line. γ can be used as a material property related parameter. By examining Eq. (10), as the contact damage propagates, the energy release rate increases, thus the change of the left term J^{*}/a can be leveled by both the increasing of J^{*} and the indentation penetration depth, a. The variation of the term in the right side of Eq. (10), $D/[a(da/dN)]$, depends on several factors. These are the indentation depth, a, the indentation speed, da/dN and D, the cyclic rate of energy associated with the damage formation. The indentation speed changes with the indentation depth. From energy balance analysis, it is clear that the value of D changes with the indentation depth, a. Thus, the variation of D is well balanced by the change in both a and da/dN. Thus, on the J^{*}/a vs $D/[a(da/dN)]$ plot, a straight line which is almost parallel to the $D/[a(da/dN)]$ axis can be obtained.

3. Experimental

The materials used include two types. One type is copper for indentation penetration zone measurement. The other one is a medium carbon steel with inclusions for simulated surface damage propagation analysis. A hardened tool carbon steel by heat treatment was used to make the wedge indenter. The indentation configuration is shown in Figure 3. The indenter has a 90° apex angle. The indentation process was conducted under cyclic loading conditions. During indentation, the load and the displacement was recorded by an Xplorer GLX data acquisition unit. These data can be used to plot and show the relation of the indentation load v.s. the nominal indenter penetration depth.

Fig. 3. Indentation set-up for performing simulated surface contact damage tests.

There exists difficulty in measuring the actual damage zone size by direct visual observation. We examined indented copper crystal using scanning electron microscopy (SEM) and measured the damage zone size. The copper polycrystal was etched in warm HCl/SnCl₄ solution. Further investigation of the damage zone using electron backscattering diffraction (EBSD) technique to reveal the contact damage zone in single crystal copper was also performed.

4. Results and discussion

The indentation cyclic load vs time is shown in Figure 4(a). Time-dependent indentation penetration depth was recorded and shown in Figure 4(b). The relation of the indentation load v.s. the indenter penetration depth at a typical cycle is shown in Figure 4(c). From the load-displacement curves, we can calculate the potential energy and the hysteresis energy associated with the contact damage processes as schematically shown in Figure 4(d). The indentation penetration depth, a, versus the number of indentation cycles, N, for three steels was plotted. The slope of the a versus N curves was used to calculate da/dN, and establish the relationship of indentation speed, da/dN, and indentation depth, a.

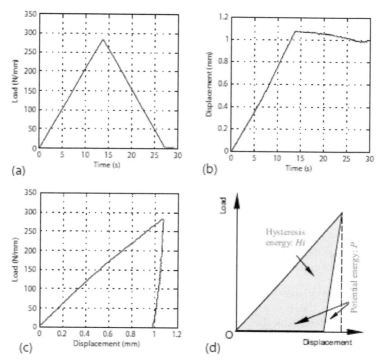

Fig. 4. Calculating energy dissipation terms from indentation test data: (a) cyclic loading profile, (b) time-dependent displacement, (c) load-displacement relationship, (d) illustration showing how to determine the potential energy and hysteresis energy.

The potential energy, P, was calculated from the loading and unloading curves recorded at intervals of number of cycles as the area above the unloading curve (see Figure 4(d)). On this basis, the relationship between the potential energy and the indentation depth, a, can be established. The relationship between P and a is used to determine the energy release rate, J^*, using Eq. (8). The hysteresis energy at each indentation cycle H_n is determined from the area of the hysteresis loop recorded as schematically shown in Figure 4(d). Based on the value of hysteresis energy and the relationship between a versus N, the quantity of D, the cyclic rate of energy dissipation into contact damage zone evolution is determined using Eq. (9).

Figure 5 shows the fatigue crack growth behavior of three medium carbon steels (named as materials A, B and C) due to the interface debonding of particle inclusions and the pearlite matrix. The carbon content of the three steels is 0.77% in weight. However, the heat treatment conditions are not the same, which affected their fatigue property. Steel A was heat treated at the highest cooling rate. B has a much lower cooling rate than A, while C has an even lower cooling rate, but close to that of B. Tension-tension fatigue tests with cyclic loading ratio of $R = 0.1$ were performed. It is found that the energy release rate and the cyclic energy dissipation rate change constantly for each of the materials during the fatigue crack growth. We also found that the critical value of energy release rate is very difficult to determine as shown in Figure 5(a), the energy release rate, J^*, versus the crack length for the three steels. The increase of the crack length, a, causes the increase of the values of J^* for the three steels. Therefore, the energy release rate can not be considered as a materials parameter for comparing the fatigue damage tolerance of different materials because a unified value for each material can not be found.

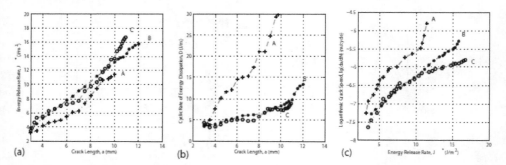

Fig. 5. The fatigue crack propagation data of three medium carbon steels: (a) the energy release rate, J^*, versus the crack length, a, (b) the cyclic rate of energy dissipation, D, versus the crack length, a, (c) crack speed versus the energy release rate.

The irreversible energy dissipation during fatigue damage of the three steels was also calculated. Based on the measured hysteresis energy for both notched and unnotched specimens and the relationship between crack length a versus fatigue cycle N, the quantity of D, the cyclic rate of energy dissipation into damage zone evolution was determined. The relationships of D and the crack length, a, for the steels, are shown in Figure 5(b). Material A displayed much higher value of the cyclic rate of energy dissipation into the active zone evolution. The other two steels, B and C demonstrated very similar behavior. For all of the three steels, it is evident that with the increase in crack length, the values of D increase.

The fatigue crack propagation speed versus the energy release rate for the three steels is shown in Figure 5(c). Steel A displayed the highest crack growth speed in the entire energy release rate range. In most part of the energy release rate range, for say, J^* less than 12 kJ/m^2, steel B and C have the crack speed very close to each other. In the energy release rate range of higher than 12 kJ/m^2, B has higher crack speed than C. It can also be seen from Figure 5(c) that the three curves display the similar two-stage crack growth behavior which are corresponding to the stable crack growth stage and the unstable crack growth stage of the specimens from the three steels. A threshold stage was observed only in the pre-crack initiation stage for A and B. But it extended to the beginning of the stable crack propagation stage for the specimens from C. In the stable crack propagation stage, the decreased acceleration in crack speed is an indicative of material damage within the area in front of the crack tip associated with fatigue crack propagation.

The damage tolerance is evaluated by the specific energy of damage γ. The parameters γ was calculated using the experimental data generated from fatigue tests including a, da/dN, J^*, and D. A plot of J^*/a versus $D/[a(da/dN)]$ can be generated for each material. Based on the results of the three steels, A, B and C, we generated Figure 6. Three straight lines which are almost parallel to the horizontal axis were obtained for the three steels. The intercepts of the three lines give the values of γ for each layer. From the results shown in Figure 6, the value of γ, being a material property related parameter, is suitable for characterizing the fatigue damage tolerance.

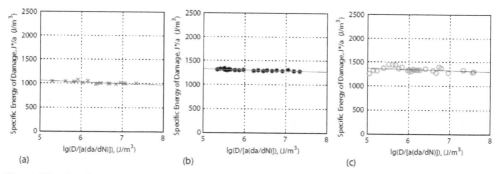

Fig. 6. Plot for determining the specific energy of damage, γ, of the three medium carbon steels under different heat treatment conditions: (a) steel A, (b) steel B, (c) steel C.

Due to the microstructure change with heat treatment conditions, the specific energy of damage for each of the steels is different. Steel A with hardening treatment, has the lowest γ, while C shows the highest γ due to tempering treatment. The specific energy of damage of steel B, heat treated at very low air cooling rate, is close to that of C. Since γ is almost a constant for each material tested, it can be taken as a parameter characteristic of the fatigue damage tolerance for evaluating the resistance to fatigue crack growth.

Although the indentation penetration depth is fairly straightforward to be recorded, it is challenge to measure the actual damage zone size. Figure 7(a) is the scanning electron microscopic (SEM) image of the copper polycrystal after etching in warm HCl/SnCl$_4$ solution. It can be seen that the grain boundaries are etched away by the solution. The precision polishing helped to expose the etching pits and islands on the surface of the specimen. These features come from the selectively dissolving of materials located near the ends of the dislocation lines. However, the indented damage zone is still unclear.

Further investigation of the damage zone using electron backscattering diffraction (EBSD) technique reveals different features within the contact damage zone. For example, the band contract map, Figure 7(b), provides the features of subgrain formation and recrystallization of the single crystal grain under wedge indentation after annealing. Since the intensity of the backscatter electrons changes from grain to grain, the grain boundary can be revealed by the band contrast change. Thus, it is possible to identify the microstructure in the area close to the indentation tip. By this method, the subgrain formation due to severe contact damage and plastic deformation can be revealed. The average size of the subgrains shown in Figure 7(b) is about 10 to 15 μm. It is also found there is an elliptical region in front of the indentation tip, which corresponds to the strain hardened elastic-plastic zone. Deeper into the indentation region, it is the fully plastic deformation zone, as shown by the in-plane lattice rotation map in Figure 7(c). Such EBSD results will provide us the insight into how to determine the size of the damage zone. For example the conservative measurement will give us the size of the damage zone the same as the indenter penetration zone (IPZ) as shown by the elliptical region in Figure 7(b). A more accurate measurement should account for the extended plastic region as shown in Figure 7(c). The distance from point A to point C or E instead of just from point O to A should be considered as the damage zone size, which is about 5 times larger than the indenter penetration zone (IPZ). This EBSD measurement results were used to correct the damage tolerance calculation by adding the contact damage zone size to the indenter penetration depth or crack length, a. Consequently, the indenter penetration speed da/dN was modified as the damage zone expansion speed.

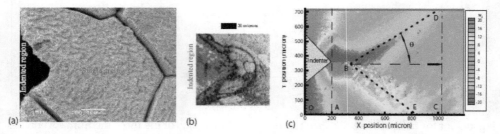

Fig. 7. Measuring the size of indentation contact damage zone via electron microscopy: (a) scanning electron microscopic measurement, (b) band contrast map of the indentation penetration zone obtained by electron backscatter diffraction (EBSD) measurement, (c) in plane lattice rotation map generated by electron backscatter diffraction (EBSD) measurement.

5. Conclusions

The energy dissipation approach is applicable for analyzing surface contact damages in various materials, including composite materials. The contact of different bodies can be modeled as indentation. Analysis of indentation and modeling of the deformation states of indented materials at different scales are performed. The stress distributions within indentation zones are described by fracture mechanics, and single crystal plasticity solutions to the stress states in regions underneath the indented zone are obtained. Instrumental indentation performed on copper materials with different grain sizes reveals both the indentation zone and damage zone. The reason for choosing copper is the high ductility of copper which allows deformation develops in a stable way during the indentation processes.

Based on the experimental studies of fatigue crack growth on three steels, the criterion for evaluating the extent of damage is identified. Although it is difficult to find a unified stress or strain based damage criterion to characterize the damage evolution, energy dissipation analysis provides a more accurate way to describe the deformation behavior of the materials. Under wedge indentation, the analysis shows advantage because the stress field has the singularity which limits the applicability of the strength criterion. The load-displacement relations with elastic-plastic responses of the materials associated with the indentation processes were obtained. The hysteresis energy was also determined. Lattice rotation measurement using electron backscatter diffraction (EBSD) technique in the region ahead of the indenter tip is an effective way to measure the dimension of the contact damage zone (CDZ) and the results can be used to define the length scales during contact deformation. A unified criterion using the hysteresis energy normalized by the length scales has been established. The above mentioned indentation tests in this work caused deformation of significant amount of materials. For further studies, comparison of deep indentation and nanoindentation should be performed.

6. References

[1] Kawaguchi, T., & Pearson, R.A. (2004). The moisture effect on the fatigue crack growth of glass particle and fiber reinforced epoxies with strong and weak bonding conditions Part 2: A microscopic study on toughening mechanism, *Composites Science and Technology*. Vol. (64): 1991-2007

[2] Kawaguchi, T. & Pearson, R.A. (2004). The moisture effect on the fatigue crack growth of glass particle and fiber reinforced epoxies with strong and weak bonding conditions Part 1: Macroscopic fatigue crack propagation behavior, *Composites Science and Technology*. Vol. (64): 1981-1989

[3] Shi, G., Zhang, M.Q., Rong, M.Z., Wetzel, B. & Friedrich, K. (2003). Friction and wear of low nanometer Si_3N_4 filled epoxy composites, *Wear*. Vol.(254): 784-796

[4] Zhang, R. & Shi, Z. (2008). Bi-interfacial debonding of coated fiber reinforced composites under fatigue load, *International Journal of Fatigue*. Vol.(30): 1074-1079

[5] Kushch, V.I., Shmegera, S.V. & Mishnaevsky, L. (2008). Meso cell model of fiber reinforced composite: Interface stress statistics and debonding paths, *International Journal of Solids and Structure*. Vol.(45): 2758-2784

[6] Cavallini, M., Di Bartolomeo, O.D. & Iacoviello, F. (2008). Fatigue crack propagation damaging micromechanisms in ductile cast irons, *Engineering Fracture Mechanics*. Vol.(75): 694-704

[7] Chan, K.S., Lee, Y.D., Nicolella, D.P., Furman, B.R., Wellinghoff, S. & Rawls, R. (2007). Improving fracture toughness of dental nanocomposites by interface engineering and micromechanics, *Engineering Fracture Mechanics*. Vol.(74):1857-1871

[8] Sharratt, B.M., Wang, L.C. & Dauskardt, R.H. (2007). Anomalous debonding behavior of a polymer/inorganic interface, *Acta Materialia*. Vol.(55): 3601-3609

[9] Shi, Z., Cui, C. & Zhou, L. (2006). Bond decay at barconcrete interface under variable fatigue loads, *European Journal of Mechanics*. Vol.(25): 808-818

[10] Kabir, M.R., Lutz, W., Zhu, K. & Schmauder, S. (2006). Fatigue modeling of short fiber reinforced composites with ductile matrix under cyclic loading, *Composite Materials Science*. Vol. (36): 361-366

[11] Okazaki, M. & Yamano, H. (2005). Mechanisms and mechanics of early growth of debonding crack in an APSed Ni-base superalloy TBCs under cyclic load, *International Journal of Fatigue*. Vol.(27):1613-1622

[12] Jiang, Y., Feng, M. & Ding, F. (2005). A reexamination of plasticity-induced crack closure in fatigue crack propagation, *International Journal of Plasticity*. Vol.(21): 1720-1740

[13] Shi, Z., Chen, Y. & Zhou, L. (2005). Micromechanical damage modeling of fiber/matrix interface under cyclic loading, *Composites Science and Technology*. Vol.(65): 1203-1210

[14] Chen, Z. Z. & Tokaji, K. (2004). Effects of particle size on fatigue crack initiation and small crack growth in SiC particulate-reinforced aluminum alloy composites, *Materials Letters*. Vol.(58): 2314-2321.

[15] Foley, M.E., Obaid, A.A., Huang, X., Tanoglu, M., Bogetti, T.A., McKnight, S.H. & Gillespie, J.W. (2002). Fiber/matrix interphase characterization using the dynamic interphase loading apparatus, *Composites A: Applied Science and Manufacturing*. Vol.(33): 1345-1348

[16] Hong, H.U., Rho, B.S. & Nam, S.W. (2002). A study on the crack initiation and growth from delta-ferrite/gamma phase interface under continuous fatigue and creep-fatigue conditions in type 304L stainless steels, *International Journal of Fatigue*. Vol.(24): 1063-1070

[17] Snodgrass, J.M., Pantelidis, D., Jenkins, M.L., Bravman, J.C. & Dauskardt, R.H. (2002). Subcritical debonding of polymer/silica interfaces under monotonic and cyclic loading, *Acta Materialia*. Vol.(50): 2395-2411

[18] Zhang, J., Wu, J. & Liu, S. (2002). Cyclically thermomechanical plasticity analysis for a broken fiber in ductile matrix composites using shear lag model, *Composites Science and Technology*. Vol.(62): 641-654

[19] Biermann, H., Kemnitzer, M. & Hartmann, O. (2001). On the temperature dependence of the fatigue and damage behaviour of a particulate-reinforced metal-matrix composite, *Materials Science and Engineering A*. Vol.(319-321): 671-674

[20] Tanaka, Y., Kagawa, Y., Liu, Y.F. & Masuda, C. (2001). Interface damage mechanism during high temperature fatigue test in SiC fiber-reinforced Ti alloy matrix composite, *Materials Science and Engineering A*. Vol.(314): 110-117

[21] Rodopoulos, C. A., Yates, J.R. & Rios, E.R., Micro-mechanical modeling of fatigue damage in titanium metal matrix composites, *Theoretical and Applied Fracture Mechanics*. Vol.(35): 59-67.

[22] Warrier, S.G., Maruyama, B., Majumdar, B.S. & Miracle, D.B. (1999). Behavior of several interfaces during fatigue crack growth in SiC/Ti-6Al-4V composites, *Materials Science and Engineering A*. Vol.(259): 189-200

[23] Foulk, J.W., Allen, D.H. & Helms, K.L.E. (1998). A model for predicting the damage and environmental degradation dependent life of SCS-6/Ti metal matrix composite, *Mechanics of Materials*. Vol.(29): 53-68

[24] Doel, T.J.A., Cardona, D.C. & Bowen, P. (1998). Fatigue crack growth in selectively reinforced titanium metal matrix composites, *International Journal of Fatigue.* Vol.(20): 35-50

[25] Warrier, S.G., Majumdar, B.S. & Miracle, D.B. (1997). Interface effects on crack deflection and bridging during fatigue crack growth of titanium matrix composites, *Acta Materialia.* Vol.(45): 4969-4980

[26] Rios, E.R., Rodopoulos, C.A. & Yates, J.R. (1997). Damage tolerant fatigue design in metal matrix composites, *International Journal of Fatigue.* Vol.(19) : 379-387

[27] Wang, P.C., Jeng, S.M., Yang, J.M. & Russ, S.M. (1996). Fatigue damage evolution and property degradation of a SCS-6/Ti-22Al-23Nb "orthorhombic" titanium aluminide composite, *Acta Materialia.* Vol.(44): 3141-3156

[28] Ghorbel, E. (1997). Interface degradation in metal-matrix composites under cyclic thermomechanical loading, *Composites Science and Technology.* Vol.(57): 1045-1056

[29] Melander, A. (1997). A Finite element study of short cracks with different inclusion types under rolling contact fatigue load, *International Journal of Fatigue.* Vol.(19): 13-24

[30] Li, C. & Ellyin, F. (1996). Fatigue damage and its localization in particulate metal matrix composites, *Matererials Science and Engineering A.* Vol.(214): 115-121

[31] Murtaza, G. & Akid, R. (1995). Modeling short fatigue-crack growth in a heat-treated low-alloy steel, *International Journal of Fatigue.* Vol.(17): 207-214

[32] Shaw, M.C., Marshall, D.B., Dalgleish, B.J., Dadkhah, M.S., He, M.Y. & Evans, A.G. (1994). Fatigue-crack growth and stress redistribution at interfaces, *Acta Metallurgica et Materialia.* Vol.(42): 4091-4099

[33] Wang, Z. & Zhang, R. J. (1994). Microscopic characteristics of fatigue-crack propagation in aluminum alloy based particulate reinforced metal matrix composites, *Acta Metallurgica et Materialia.* Vol.(42): 1433-1445

[34] Gradin, P.A. & BÄacklund, J. (1981). Fatigue debonding in fibrous composites, *International Journal of Adhesion and Adhesives.* Vol.(1): 154-158

[35] Horst, J.J., Salienko, N.V., Spoormaker, J.L. (1998) Fibre-matrix debonding stress analysis for short fibre-reinforced materials with matrix plasticity, finite element modelling and experimental verification, *Composites A: Applied Science and Manufacturing.* Vol.(29): 525-531

[36] Horst, J.J. & Spoormaker, J.L. (1996). Mechanisms of fatigue in short glass fiber reinforced polyamide 6, *Polymer Engineering Science.* Vol.(36): 2718-2726

[37] Horst, J.J. & Spoormaker, J.L. (1997). Fatigue fracture mechanisms and fractography of short glass fiber reinforced polyamide 6, *Journal of Materials Science.* Vol.(32): 3641-3651

[38] Botsis, J. & Zhao, D. (1997). Fatigue fracture process in a model composite, *Composites A: Applied Science and Manufacturing.* Vol.(28): 657-666

[39] Dessureautt, M. & Spelt, J.K. (1997). Observations of fatigue crack initiation and propagation in an epoxy adhesive, *International Journal of Adhension and Adnesives.* Vol.(17): 183-195

[40] Kysar, J.W., Gan, Y.X. & Mendez-Arzuza, G. (2005). Cylindrical void in a rigid-ideally plastic single crystal Part I: Anisotropic slip line theory solution for face-centered cubic crystals, *International Journal of Plasticity.* Vol.(21): 1481-1520

[41] Aglan, H., Gan, Y.X., Chu, F. & Zhong, W.H. (2003). Fatigue damage analysis of particulate and fiber filled PTFEs based on energy expended on damage formation, *Journal of Reinforced Plastics and Composites*. Vol.(22): 339-360

[42] Aglan, H. & Gan, Y.X. (2001). Fatigue crack growth analysis of a premium rail steel, *Journal of Materials Science*. Vol.(36): 389-397

[43] Aglan, H.A., Gan, Y.X., Chin, B.A. & Grossbeck, M.L. (2000). Effect of Composition on the fatigue failure behavior of vanadium alloys, *Journal of Nuclear Materials*. Vol.(278): 186-194

[44] Aglan, H., Gan, Y.X., Chin, B.A. & Grossbeck, M.L. (1999). Fatigue failure analysis of V-4Ti-4Cr alloy, *Journal of Nuclear Materials*. Vol.(237): 192-202

[45] Gan, Y., El-Hadik, M., Aglan, H., Faughnan, P. & Bryan, C. (1999) Fatigue crack growth analysis of PCTFE, *Journal of Elastomers and Plastics*. Vol.(34): 96-129

[46] Aglan, H., Gan, Y., El-Hadik, M., Faughnan, P. & Bryan, C. (1999). Evaluation of fatigue fracture resistance of unfilled and filled PTFE materials, *Journal of Materials Science*. Vol.(34): 83-97

8

Progressive Stiffness Loss Analysis of Symmetric Laminated Plates due to Transverse Cracks Using the MLGFM

Roberto Dalledone Machado, Antonio Tassini Jr.,
Marcelo Pinto da Silva and Renato Barbieri
Pontifical Catholic University of Parana, Mechanical Engineering Graduate Program,
Brazil

1. Introduction

With the purpose to create high strength advanced structures, new materials are being developed presenting favorable characteristics for specific applications. Composite Materials are examples of these developments. They can be formed by high strength long fibers, conveniently oriented in a matrix, to form a lamina of composite material. The lamina presents high strength in the fiber direction, but, since it is slender, does not have enough rigidity, what makes impossible the use of an isolated lamina. Piling up and gluing a set of laminas, a laminate is formed which one presents better characteristics than original and isolated materials. The main strength of each lamina is oriented according to the fiber directions. Thus, micro cracks can be produced if sufficient tension is applied in the transverse direction of the fibers, as shown in figure 1, since the resistance of the lamina in these directions depends only on the matrix material. The rise of several transverse cracks produces loss of stiffness in the laminate.

Several papers are found in technical literature dealing with the behavior of composite materials with transverse cracks. Vejen & Pyrz (2002) investigated the transverse crack growth in long fiber composites using the finite element method. Three criteria concerning pure matrix growth, fiber/matrix interface growth and crack kinking out of a fiber/matrix interface were implemented to form a software package for crack propagation calculus.

Cain and colleagues (2003) have studied unidirectional graphite bismaleimide composites to determine the effect of the matrix dominant properties on the failure of the material. The authors showed that the final fracture was caused by the development of a dominant matrix shear crack parallel to the fibers. They also concluded that the decrease in shear modulus of the composite was the most sensitive and best represented by damage evolutions.

Ogihara et al (1998) have proposed a two-dimensional model which considers that, in the case of displacements and stress fields in the interlaminar cross-play laminates, there is a prevalence of plane-strain case, even in the presence of transverse cracks. They have also commented that the failure process of cross-ply laminates is due an accumulation of transverse cracks and delamination.

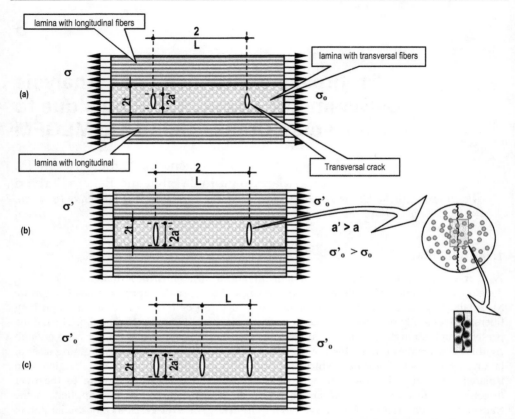

Fig. 1. (a) A composite $[0\degree/90\degree/0\degree]_s$ laminate plate with transverse micro-cracks in the matrix; (b) extension of pre-micro cracks; (c) formation of new micro crack.

An analytical model based on the principle of minimum potential energy was developed by Ji *et al* (1998) and applied to determine the two-dimensional thermoelastic stress state in cross-ply composite laminates containing multiple equally spaced transverse cracks in the 90° plies subjected to tensile loading in the longitudinal direction. The criterion of strain energy release rate was employed to evaluate the critical applied stresses for two of the possible fracture modes. After some numerical experiments, they have concluded that the formation of new cracks never takes place until pre-existing cracks extends through the entire thickness of the 90° plies.

Wada *et al* (1999) have presented a damage mechanics model to predict the nonlinear behavior of laminated composites due to crack evolution. A new concept of cracking layer is proposed by a technique based on uniform work-softening layer. With this concept, the constitutive equations for a cracking layer are constructed according to modern plasticity theory. So, the lamina damage surface is defined in the stress space and the constitutive equations for a cracking layer are constructed by applying the defined damage surface to the associated flow rule.

One of the first damages that occurs in composite laminate are the transverse cracks, as mentioned by Allen and colleagues (Allen et al 1987 a,b) and Tay and Lim (Tay & Lim, 1996). The cracks appear in a layer where the highest stress values act transversally to the fiber, exceeding the matrix resistance. With loading increment, the increase on the number of transverse cracks may happen in a diffuse way reducing the structural rigidity. The accumulation of this damage can accelerate the beginning of delamination, changing the natural frequency of the composite structure and causing a greater degradation in severe environment, jeopardizing its service life.

After the initiation and development of micro cracks, there is a process of accumulation of damage that reduces the structural stiffness.The tolerance to the damage is related to the stiffness of the structure which, in turn, is affected by the accumulation of micro defects during loading. The process of damage evolution in composite laminate is generally very complex due to the multiplicity of failure modes such as transverse cracks, delamination, decoupling fiber-matrix interface, and fiber breakage. The characterization of this process is generally possible when single cases are analyzed, where each failure mode can be separated and studied individually. The use of Fracture Mechanics, especially in terms of linear elastic fracture, has presented good results for isotropic material because, in this case, can be adequately characterized by a single parameter (the stress intensity factor). However, attempts to apply this method in composite laminates, whose behavior is orthotropic, have met unsatisfactory results, mainly when transverse cracks in the matrix are studied. Therefore, to determine changes of the mechanical properties in a laminate, the total number of cracks formed in the transverse layers must be taken to account, or, under a generalized crack distribution, the most appropriated methodology is based on Damage Mechanics.

Many researchers have developed studies to evaluate the properties of laminates subjected to generalized cracks in the matrix. Among these ones, can be cited the papers of Allen et. al. (1987 a,b), Hashin (1987), Talreja (1984) and Lim & Tay.

The present paper has the objective to apply the Continuum Damage Mechanics Theory to long fiber laminate composites. The transverse cracks appearance in the matrix implies in a rigidity loss due to damage accumulation. The increase of the load is considerate monotonically. Several failure criterions are presented and implemented such as, the Maximum Stress Criterion, the Maximum Strain Criterion, Tsai-Hill and Tsai-Wu Criterion. The proposed methodology is restrict to the case of symmetric laminate and it is evaluated by a numerical approximation technique known as Modified Local Green's Function Method (MLGFM), which one will be briefly descript on this paper.

2. Representation of the generalized damage in symmetric laminates

2.1 Constitutive relations

The models developed by Talreja & Boehler (1990), Allen et. al. (1987 a,b) and Lim and Tay (1996) to describe the damaged composite laminates were based on the Continuum Damage Mechanics using internal state variables. In the presente paper, the model proposed by Allen et al (1987 a,b) will be used, which describes the damage through a set of internal state variables. The final result of the distributed damage is built in the constitutive equations through these variables. Thus, the stress-strain relationship of the representative volume of a damaged material at the level of a lamina is assumed as:

$$\sigma_{ij} = C_{ijkl}\varepsilon_{kl} + I^{\eta}_{ijkl}\alpha^{\eta}_{kl} \tag{1}$$

where σ_{ij} is the applied stress tensor, C_{ijkl} is the constitutive relation tensor of the undamaged material, ε_{kl} are the strain tensor, I^{η}_{ijkl} are the elements of the damage matrix, α^{η}_{kl} are the internal state variables, and $\eta = 1, 2, 3, \ldots$, refers to the damage modes. As suggested by Allen *et al* (1987 a,b), a first simplification can be made considering that the tensor I_{ijkl} is the actual tensor of constitutive relationships, as shown in Equation (2).

$$I^{\eta}_{ijkl} = C_{ijkl} \tag{2}$$

However, it is important to emphasize that equation (1) does not provide any information on how the damage state has been attained, that is, the history of damage accumulation. Thus, it is necessary to turn to Fracture Mechanics in search of a suitable criterion to evaluate the damage growth. Thereby, equation (1) is sufficiently general to permit the use of Classical Laminate Theory to determine the composite laminate constitutive relations with transverse cracks in the matrix.

Supposing the representation of the laminated plate by plane elements located in its middle surface, the loads in a certain point inside this surface can be evaluated by the following expressions:

$$\{N\} = \int_{-t/2}^{t/2}\{\sigma_x \quad \sigma_y \quad \tau_{xy}\}dz \tag{3}$$

$$\{M\} = \int_{-t/2}^{t/2}\{\sigma_x \quad \sigma_y \quad \tau_{xy}\}zdz \tag{4}$$

where $\{N\}$ e $\{M\}$ are, respectively, the force and moment resultants vectors, σ_x, σ_y, σ_{xy} are the stresses in the plane of the lamina and t is the thickness of the laminate. Taking to account that $\{\varepsilon_0\}$ e $\{\kappa_0\}$ are the strain and bending vectors in the middle surface of the plate, $[A]$, $[B]$ and $[D]$ are the laminate extensional stiffness matrix, coupling stiffness matrix and bending stiffness matrix, respectively, $\{D^N\}$e $\{D^M\}$ are the damage vectors related to the force and moment resultants, the expressions (3) and (4) can be transformed to:

$$\{N\} = [A]\{\varepsilon_0\} + [B]\{\kappa_0\} + \{D^N\} \tag{5}$$

$$\{M\} = [B]\{\varepsilon_0\} + [D]\{\kappa_0\} + \{D^M\} \tag{6}$$

Assuming that z_{k-1} e z_k are the corresponding distances from the middle surface to the inner and outer surfaces of the k_{th} lamina, respectively, $[\bar{Q}]_k$ and $\{\bar{\alpha}\}_k$ are the transformed reduced material stiffness matrix and the transformed vector of the internal state variables (expressed in global coordinates), respectively. The matrix and vectors presents in equations (5) and (6) can be expressed by:

$$[A] = \sum_{k=1}^{N}(z_k - z_{k-1})\left[\bar{Q}\right]_k \tag{7}$$

$$[B] = \frac{1}{2}\sum_{k=1}^{N}\left(z_k^2 - z_{k-1}^2\right)\left[\bar{Q}\right]_k \tag{8}$$

$$[D] = \frac{1}{3}\sum_{k=1}^{N}\left(z_k^3 - z_{k-1}^3\right)\left[\bar{Q}\right]_k \tag{9}$$

$$\{\varepsilon^0\} = \int_{-t/2}^{t/2}\{\varepsilon_x \quad \varepsilon_y \quad \gamma_{xy}\}dz \tag{10}$$

$$\{\kappa^0\} = \int_{-t/2}^{t/2}\{\kappa_x \quad \kappa_y \quad \kappa_{xy}\}dz \tag{11}$$

$$\{D^N\} = \sum_{k=1}^{N}\left(z_k - z_{k-1}\right)\left[\bar{Q}\right]_k\{\bar{\alpha}\}_k \tag{12}$$

$$\{D^M\} = \frac{1}{2}\sum_{k=1}^{N}\left(z_k^2 - z_{k-1}^2\right)\left[\bar{Q}\right]_k\{\bar{\alpha}\}_k \tag{13}$$

The internal state variables vector has two components and is expressed by:

$$\{\alpha\}_k = \sum_{\eta=1}^{P}\{0 \quad \alpha_{22}^{\eta} \quad \alpha_{12}^{\eta}\}_k \tag{14}$$

where P is the total number of damage models being considered and α_{22} and α_{12} are the internal state variable of the problem.

2.2 Determination of internal variables

In spite of the random characteristic, as can be found in the work of Silberschmidt (2005), the transverse cracks are assumed to be uniformly distributed. In this way, the laminate behavior can be adequately represented by a representative unit volume of material containing a transverse crack, as shown in figure 2. In the particular case of symmetric laminates, the damage models are simplified and incorporate only two types of fracture, namely, Mode I (crack opening) and Mode I coupled with Mode III (shear out of plane). They are represented respectively by the internal variables α_{22} e α_{12}. As only symmetric laminates are analyzed in this paper, just the a_{22} variable will be developed.

The internal variable, equation (15), proposed by Allen (Allen et al., 1987 a,b) can be determined by a computational analysis based on Finite Element Method. The representative volume is modeled, as shown in figure 3, for the symmetric laminate [0°/90°/0°]. A uniform displacement is imposed in one side of the element to determine the opening of the crack. The size and shape of the representative volume depend on the thickness of the different plies and the crack density (the number of cracks per unit of volume). Then, the internal variable can be determined by:

$$\alpha_{22} = \frac{1}{V}\int_{S_c} u_2 n_2 dS \tag{15}$$

where u_2 is the crack opening displacement, n_2 is the unitary vector normal to the crack surface, V is the representative element volume and S_c is the crack surface.

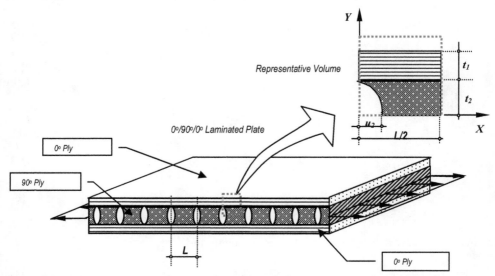

Fig. 2. A [0°/90°/0°] laminated plate with generalized cracks: definition of parameters and the representative volume (Machado et al, 2008).

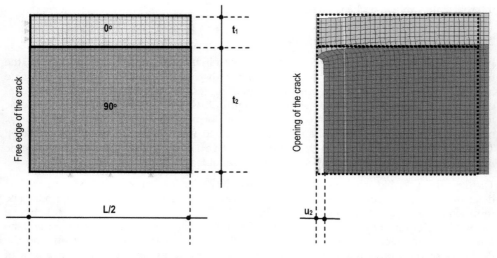

Fig. 3. Boundary conditions and finite element mesh to evaluate the crack opening in a representative volume [0°/90°/0°] – (as suggested by Lim & Tay, 1996)

Considering that t_1 and t_2 are the thickness of the 0° and 90° plies, respectively, t is the total thickness of the laminate, l is the distance between two adjacent transverse cracks, ρ is the non-dimensionalized crack density (the quantity of cracks per unit of length), δ is the non-dimensionalized maximum crack opening displacement, u_2 is the maximum crack opening displacement, ψ is a normalized function of the crack opening profile and ξ is the normalized distance between the cracks center, the following expressions can be defined:

$$t = t_1 + t_2 \tag{16}$$

$$\rho = \frac{2t}{L} \tag{17}$$

$$\delta = \frac{u_2}{t_2} \tag{18}$$

$$\psi = \frac{u(\xi)}{u_2} \tag{19}$$

The maximum crack opening displacement u_2 can be determined by a simple finite element analysis, considering the boundary conditions specified in figure 3. An arbitrary displacement is imposed. The value of δ is determined by the equation (18), and the displacement u_2 is determined by MEF. The non-dimensionalized maximum crack opening displacement δ can be obtained using ρ, as shown in the expression (20):

$$\delta = c_1 \left(e^{-a_1 \rho} \right) + c_2 \left(e^{-b_1 \rho} \right) + c_3 \tag{20}$$

The constants a_1, b_1, c_1, c_2 e c_3 in the expression (20) depends on the type of the used material. The table 1 exposes the value of these constants for the laminate glass/epoxy (Gl/Ep).

Material	Formulation in terms of ρ				
	c_1	c_2	c_3	a_1	b_1
Glass/Epóxi	1.03	-0.81	2.28E-2	0.94	1.00

Table 1. Coefficients for the expression (20) (Lim & Tay, 1996)

As the internal variable used in this problem depends on the maximum crack opening displacement according to equation (15), and the crack density is calculated by $\zeta = 1/L$, it can be shown that the state variable associated to the Mode I becomes:

$$\text{Mode I:} \qquad \alpha_{22} = \frac{8}{5} u_2 \zeta \tag{21}$$

3. Approximation by computational methods

In this paper, the maximum crack opening displacement u_2 is determined by the expressions (18) and (20). The crack density ρ depends on the distance between two adjacent transverse cracks l, and its values are successively modified by the verification of the composite material rigidity loss.

Considering a conventional structural approximation by conventional Finite Element Method, the problem can be expressed by a system of algebraic equations, representing a typical element:

$$
\begin{bmatrix} K_{11} & K_{12} & K_{16} \\ K_{21} & K_{22} & K_{26} \\ K_{61} & K_{62} & K_{66} \end{bmatrix} \begin{Bmatrix} d_x \\ d_y \\ d_z \end{Bmatrix} = \begin{Bmatrix} F_1^a \\ F_2^a \\ F_6^a \end{Bmatrix} + \begin{Bmatrix} F_1^d \\ F_2^d \\ F_6^d \end{Bmatrix}
\tag{22}
$$

where K_{ij} are the element stiffness matrix, (d_x, d_y, d_z) are the components of the element displacement vector and $\{F^a\}$ and $\{F^d\}$ are the applied force vector and the element damage force vector.

The procedure used in this paper to obtain the expected results is a little different because it uses a different computational method known as Modified Local Green's Function Method (MLGFM), in witch the system defined in expression (22) is not directly applied in a conventional FEM. A detailed explanation of this procedure can be found in Barbieri *et al* (1998a,b) and Machado *et al* (2008). The MLGFM is an integral method that determines the unknowns on the boundary, similarly to the Boundary Element Method, but the fundamental solutions are generated automatically by projections of the Green's Functions developed from de field, as in the Finite Element Method. The matrix and the vectors indicated in (22) will be used to produce values at the boundary, as explained in the next topic.

3.1 The Modified Local Green's Function Method - MLGFM

The Modified Local Green's Function Method (MLGFM) is an integral technique that associates the Finite Element Method and the Boundary Element Method, solving the problem through an integral equations system. Unlike to the BEM and the Trefitz Methods, the MLGFM does not use a fundamental solution and/or a Green's function. The term "Local" indicates that the calculation of the GF projections can be done locally, that is, for each element.

Essentially, the MLGFM uses a transverse integration technique and reciprocity relations to determine, at a local level, the Green's Function, transforming the partial differential operator in an ordinary partial operator (Barcelos & Silva, 1987). The MLGFM uses finite elements at the domain with the purpose to create discrete projections of the Green's Function, corresponding to fundamental solutions, that are used later in the integral equations system associated to the boundary approximation. To analyze a continuum mechanics problem through the MLGFM, such as the plate indicated in figure 4, two meshes are necessary, one for the domain and other for the boundary. With domain elements, the method generates a set of domain equations, which are used to generate automatically the domain Green's projections. Later, a set of boundary equations are also generated and the boundary Green's projections can be determined with the domain projections developed before. At the end, the system is solved only for boundary equations, where the main variables are calculated. Domain values may be obtained once the boundary values are known after the solution of the boundary equation system.

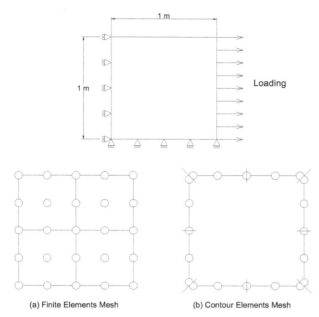

(a) Finite Elements Mesh   (b) Contour Elements Mesh

Fig. 4. Symmetric boundary conditions and plate for a 2x2 mesh (4 finite elements and 8 contour elements) by the MLGFM.

The most important steps of the MLGFM are detailed in the work sof Barbieri *et al* (1998a) and Machado *et al* (2008). It is possible to show that through the MLGFM two sets of equations are formed, the first one in the domain (equation (23)) and the other one on the boundary (equation (24)):

$$\mathbf{u}(Q) = \int_{\Omega} [\mathbf{G}^{\mathrm{T}}(P,Q) \, \mathbf{a}(P)] d\Omega + \int_{\Gamma} [\mathbf{G}^{\mathrm{T}}(p,Q) \, \mathbf{f}(p)] d\Gamma \ ; \ P,Q \in \Omega \ ; p \in \Gamma \tag{23}$$

$$\mathbf{u}(q) = \int_{\Omega} [\mathbf{G}^{\mathrm{T}}(P,q) \, \mathbf{a}(P)] d\Omega + \int_{\Gamma} [\mathbf{G}^{\mathrm{T}}(p,q) \, \mathbf{f}(p)] d\Gamma \ ; \ P \in \Omega \ ; p, q \in \Gamma \tag{24}$$

where Q, P are two points in the domain; q, p are other two points on the boundary; $\mathbf{a}(P)$ is the vector of independent terms for the original problem; $f(p)$ is the vector associated to the fluxes on the boundary; $\mathbf{G}(i,j)$ are the Green's functions which may be understood as the generalized displacement in the point i in the direction of an unitary vector $\mathbf{n_i}$, when a generalized force is applied over the point j, in the direction of a unitary vector $\mathbf{n_j}$.

Equations (23) and (24) describe completely the problem. Since these equations involve domain and boundary integrals, two types of meshes are necessary, one in the domain and the other on the boundary, using FE and BE methods, respectively. The FE domain approximation is also used to develop the Green's functions which are associated to the matrices $\mathbf{G}(P,Q)$, $\mathbf{G}(p,Q)$, $\mathbf{G}(P,q)$ and $\mathbf{G}(p,q)$.

The approximation shape functions are the same as in the conventional FE and BE methods. For the present work, quadratic shape functions were employed to construct nine nodes lagrangean finite elements and three nodes boundary elements.

Developing discrete equations from the nodal values, two sets of linear equations are determined:

$$\mathbf{A}\, \mathbf{u}_\Omega = \mathbf{B}\, \mathbf{f} + \mathbf{C}\, \mathbf{a} \quad \text{(in the domain)} \tag{25}$$

$$\mathbf{D}\, \mathbf{u}_\Gamma = \mathbf{E}\, \mathbf{f} + \mathbf{F}\, \mathbf{a} \quad \text{(on the boundary)} \tag{26}$$

where $\mathbf{u}_\Omega$ and $\mathbf{u}_\Gamma$ are the domain and the boundary displacements, respectively, $\mathbf{a}$ and $\mathbf{f}$ are the independent and the fluxes variables vectors. The matrices $\mathbf{A}$, $\mathbf{B}$, $\mathbf{C}$, $\mathbf{D}$, $\mathbf{E}$ and $\mathbf{F}$ can be written as:

$$\mathbf{A} = \int_\Omega \psi(Q)^T \psi(Q) d\Omega \tag{27}$$

$$\mathbf{B} = \int_\Omega \psi(Q)^T\, G_\Gamma(Q) d\Omega \tag{28}$$

$$\mathbf{C} = \int_\Omega \psi(Q)^T\, G_\Omega(Q) d\Omega \tag{29}$$

$$\mathbf{D} = \int_\Gamma \phi(q)^T \phi(q) d\Gamma \tag{30}$$

$$\mathbf{E} = \int_\Gamma \phi(q)^T\, G_\Gamma(q) d\Gamma \tag{31}$$

$$\mathbf{F} = \int_\Gamma \phi(q)^T\, G_\Omega(q) d\Gamma \tag{32}$$

where $\psi(Q)$ and $\phi(q)$ are matrices with the shape functions in the domain and on the boundary, respectively, and $G_\Gamma(Q)$, $G_\Omega(Q)$, $G_\Gamma(q)$, $G_\Omega(q)$ are the Green's function projections over the boundary Γ and the domain Ω, evaluated on the points Q and q. The Green's projections can be written as:

$$G_\Gamma(Q) = \int_\Gamma G^T(p,Q)\, \phi(p)\, d\Gamma \tag{33}$$

$$G_\Omega(Q) = \int_\Omega G^T(P,Q)\, \psi(P)\, d\Omega \tag{34}$$

$$G_\Gamma(q) = \int_\Gamma G^T(p,q)\, \phi(p)\, d\Gamma \tag{35}$$

$$G_\Omega(q) = \int_\Omega G^T(P,q)\, \psi(P)\, d\Omega \tag{36}$$

In order to determine the Green's functions automatically, it must be considered the following functional **F**, which one depends on G_Ω or G_Γ:

$$F(G_\Omega,G_\Gamma) = B(G,G) - \alpha\, B_1(G_\Omega,\psi) - \beta\, B_2(G_\Gamma, \phi) + B_3(G,G) \tag{37}$$

where

G – corresponds to G_Ω or G_Γ depending on the case of interest;
B – is a bilinear form, developed to G_Ω or G_Γ;
α and β are constants whose values are:
    $\alpha = 1$ and $\beta = 0$ to determine G_Ω
    $\alpha = 0$ and $\beta = 1$ to determine G_Γ
B_1, B_2, B_3 – are bilinear forms which can be written as:

$$B_1(G_\Omega,[\psi]) = \int_\Omega G_\Omega(Q)\, \psi(Q)\, d\Omega \tag{38}$$

$$B_2(G_\Gamma,[\phi]) = \int_\Gamma G_\Gamma(q)\, \phi(q)\, d\Gamma \tag{39}$$

$$B_3(G,G) = \tfrac{1}{2}\int_\Gamma N^{\#}(q)\, G\,(q).\, G\,(q)\, d\Gamma \tag{40}$$

As in the variational approach of the FEM, the minimization of functional $F(G_\Omega,G_\Gamma)$ in Equation (37) results in a linear equation system which can be solved to determine the Green's projections

$$[K][G_\Omega(Q) \;\vdots\; G_\Gamma(Q)] = [A \;\vdots\; D] \tag{41}$$

where [K] is the global stiffness matrix, evaluated in the same way of the conventional finite element stiffness matrix; **A** and **D** are the matrices of Equations (27) and (30), respectively. In this way, the Green's projections are determined directly from Equation (41), and can be applied in Equations (28), (29), (31) and (32) to complete the matrices of equations (25) and (26), which are the main system of the MLGFM.

4. Damage evolution

With purpose to quantify the damage accumulation due to a monotonic load increment, some failure criterion will be used. Generically, failure criteria can be considered as:

$$F_a\left(\frac{\sigma_{ij}}{X_{ij}}\right) = Z \tag{42}$$

where F_a is the failure criterion, σ_{ij} are the local stress, X_{ij} are the principal material strength and Z is the failure value characteristic to each criterion. The rigidity degradation of a component occurs due to a progressive process during its serviceable life. It is important to note that the evolution of rigidity loss in a structure can be characterized by a single crack or by the occurrence of generalized cracks. Combinations of failure modes can act together causing changes in the material properties and in its local stress distribution. In this way, the main difficulty in this kind of analysis is the adoption of a failure criterion, F_a, that conveniently describes the damage evolution due to a failure mode.

The theories introduced to prevent the failure of an orthotropic laminate are adaptations of failure criterion for isotropic materials, modified for biaxial stress cases, such as, Maximum Stress Criterion, Maximum Strain Criterion, Tsai-Hill Criterion and Tsai-Wu Criterion (Reddy, 1997; Vasiliev & Morozov, 2001; Mendonça, 2005).

A new criterion is presented, based on the strain energy release rate, to evaluate the formation of a new micro crack (Anderssen et al., 1998; Ji et al., 1998; Kobayashi et al.,2000). The released energy is used because it is practically independent from the crack length (Anderssen et al., 1998). Some of these criterions are presented here.

4.1 Maximum stress criterion

According to the Maximum Stress Criterion, for orthotropic materials, while the stresses in the principal directions of the material are lower than strength of the material in this direction, there are no fails, which means:

Tensile failure
$$\sigma_1 \leq X_t \text{ - longitudinal direction}$$
$$\sigma_2 \leq Y_t \text{ - transverse direction}$$

Compressive failure
$$\sigma_1 \leq X_c \text{ - longitudinal direction}$$
$$\sigma_2 \leq Y_c \text{ - transverse direction}$$

Shear failure
$$\tau_{12} \leq C \text{ - plane shear}$$

where X_t is the longitudinal tensile strength, Y_t is the transverse tensile strength, X_c is the longitudinal compressive strength, Y_c is the transverse compressive strength and C is the shear strength of the lamina.

4.2 Maximum strain criterion

This theory is analogous to the Maximum Stress Criterion, but the fail criterion is controlled by deformation limits in the principal directions of material. In this theory the material will fail when one of the following limits are reached:

Tensile failure
$$\varepsilon_1 \leq X_{\varepsilon t} \text{ - longitudinal direction}$$
$$\varepsilon_2 \leq Y_{\varepsilon t} \text{ - transverse direction}$$

Compressive failure
$$\varepsilon_1 \leq X_{\varepsilon c} \text{ - longitudinal direction}$$
$$\varepsilon_2 \leq Y_{\varepsilon c} \text{ - transverse direction}$$

Shear failure
$$\gamma_{12} \leq C_{\varepsilon} \text{ - plane shear}$$

where X, X, Y, Y are the maximum deformation deformation values in the principal directions 1 and 2, for tensile and compressive loading, C_ε, is the maximum angular distortion in the plane 1-2.

4.3 Tsai-Hill Criterion

An adaptation made by Tsai in the Hill Criterion for transverse orthotropic laminate at plane stress condition, resulted in the expression (43):

$$\frac{\sigma_1^2}{X^2} + \frac{\sigma_2^2}{Y^2} - \frac{\sigma_1\sigma_2}{X^2} + \frac{\tau_{12}^2}{C^2} = 1 \tag{43}$$

4.4 Tsai-Wu Criterion

A simple procedure was proposed by Tsai-Wu, changing the Tsai-Hill Criterion in equation (43). When the tensile and compression strength are similar, the expression (44) becomes the Tsai-Hill Criterion.

$$\frac{\sigma_1^2}{X^2} + \frac{\sigma_2^2}{Y^2} - \frac{\sigma_1\sigma_2}{XY} + \frac{\tau_{12}^2}{C^2} = 1 \tag{44}$$

4.5 Strain Energy Release Rate Criterion

The Strain Energy Release Rate Criterion to describe the damage evolution was applied by Lim and Tay (Lim & Tay, 1996). Consider a symmetric laminated composite of width b and length L, with the configuration $[0°_l/90°_m]_s$, where l and m are integers. When the laminated is loaded uniaxially in tension, the stress-strain curve is linear until the failure criterion is reached for the first time, at point A (figure 5). A transverse crack is introduced in 90° layer.

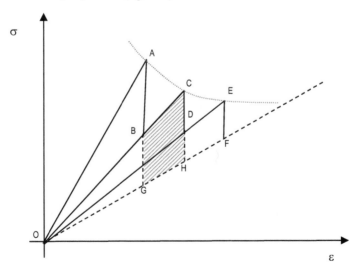

Fig. 5. Stiffness loss in composite laminates $[0°_l/90°_m]_s$ – Lim & Tay (1995).

The result is a reduction in the effective stiffness of the laminate in the loading direction, and this is represented by the OB segment in figure 5. Upon further loading, this reduction is verified by the segment BC. This process is repeated until line OF is reached. Note that this dotted line represents the stiffness of the laminate where the contribution of the 90° layers was neglected.

When the area BCHG reaches a critical value. An additional transverse crack is formed and the effective stiffness reduces again, as indicated by the segment OC. Denoting the area BCHG in figure 5 as U_{0i}, where i indicates the lamina in analysis, the strain energy density is given by:

$$U_{0i} = \frac{1}{2}\left(\sigma_{xi}\Delta\varepsilon_{xi} + \sigma_{yi}\Delta\varepsilon_{yi} + \tau_{xyi}\Delta\gamma_{xyi}\right) \tag{45}$$

Where σ_{xi}, σ_{yi} e τ_{xyi} are the stress in x direction, y direction and xy plane shear of the lamina i, respectively, and $\Delta\varepsilon_{xi}$, $\Delta\varepsilon_{yi}$ e $\Delta\gamma_{xyi}$ are the strain in x direction, y direction and xy plane of the lamina i, respectively. Therefore, the energy U_i, necessary to form a new crack, can be defined as:

$$U_i = \frac{t}{t_2}LU_{0i} \tag{46}$$

Where t is the thickness of the laminated, t_2 is the thickness of the 90° plies an L is the length of the laminated.

In this way, a transverse crack is assumed to be formed when:

$$U_i \geq G_{Ic} \tag{47}$$

Where G_{Ic} is the mode I energy release rate for the formation of a transverse crack. The process of determining the transverse crack density is repeated for each successive micro crack, using the same value for G_{Ic}. As seen in figure 5, a series of points (A, B, C, D, ..., E) can be generated until the limit OI is reached. From this limit, matrix cracking in the 90° layers no longer influences significantly the laminate stress-strain behavior. It must be observed that in practice, the intervals between the points are very small, turning the curve smooth, rather than the curve shown in figure 5.

5. Applications

5.1 Analysis of laminated plates by the MFLGM

The first application refers to the analysis of a laminate plate, whose materials of its lamina are defined in table 2. The aim of this application is to determine the stiffness loss E/Eo due to the improvement of crack density ζ for the Gr/Ep [0°./90°]s laminated, using $\delta(\rho)$ formulation

Material	E_{11} (GPa)	E_{22} (GPa)	G_{12} (GPa)	G_{23} (GPa)	υ_{12}
Grafite / Epoxi (Gr/Ep)	142,00	9,85	4,48	3,37	0,3
Glass / Epoxi (Gl/Ep)	41,70	13,00	3,40	3,40	0,3

Table 2. Material Properties - Highsmith e Reifsnider (1982)

The loss of stiffness is observed in figure 6 for different meshes and compared with the results obtained by Lim & Tay (1996) and experimental results. As the crack density grows up, the stiffness diminishes. The results are better with finest meshes, but even with coarse meshes, the approximation is good. Figure 7 shows the loss E/Eo versus crack density ζ for the case Gl/Ep $[0°/90°]_s$ – using $\delta(\theta)$ formulation. The same considerations are made for this case.

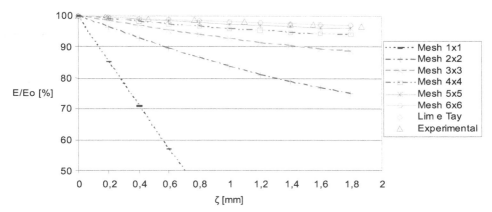

Fig. 6. Stiffness loss E/Eo versus crack density ζ for the Gr/Ep $[0°./90°]$s laminated, using $\delta(\rho)$ formulation.

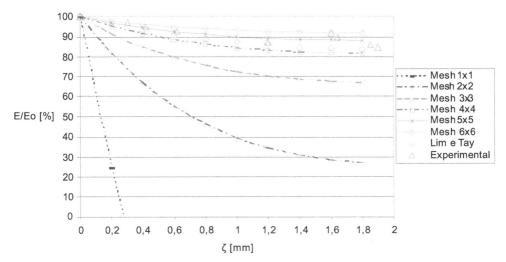

Fig. 7. Stiffness loss E/Eo versus crack density ζ for the case Gl/Ep $[0°/90°]_s$ – using $\delta(\theta)$ formulation

5.2 Progressive stiffness loss of laminate

To evaluate the stiffness loss of laminated plates due to micro-crack accumulation under increasing monotonic loading using the MLGFM, the following conditions were considered:

a. Stress-strain relations of a thin orthotropic laminate are considered in plain stress state;
b. Dimensions of the squared plate are 2,0 m x 2,0 m, but only a ¼ was modeled due to its double symmetry: $\{(x,y) \in R^2 : (0 < x < 1,0; 0 < y < 1,0\}$;
c. Axial tension loading in "x" direction;
d. The properties of the material used are listed in the tables 2 and 3;
e. The value of G_{Ic} adopted is 250 J/m² for the glass/epoxi laminate (Tay & Lim, 1993).

Material	E_{11} (GPa)	E_{22} (GPa)	G_{12} (GPa)	G_{23} (GPa)	v_{12}
Glass / Epoxy (Gl/Ep)	41.70	13.00	3.40	3.40	0.3

Table 3. Mechanical properties (Highsmith & Reifsnider, 1982)

X_t (MPa)	Y_t (MPa)	X_c (MPa)	Y_c (MPa)	C (MPa)
1170.00	32.00	53.00	18.00	45.00

Table 4. Strength limits for glass/epoxy laminate (Highsmith & Reifsnider, 1982)

In order to compare the failure criterion, a $[0º/90º_3]_s$ glass/epoxy symmetric laminated with total thickness of 1,624mm was used. All layers on the laminated have the same thickness. The results are presented in figure 8. All criterions were implemented in the same program to facilitate the comparison.

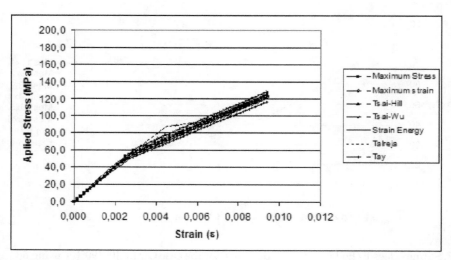

Fig. 8. Gl/Ep $[0º/90º_3]_s$ Laminated – Failure Criterion comparison.

To demonstrate the model capability to prevent the laminate stiffness loss, a comparison between the results obtained by the strain energy criterion implemented in this paper and the results obtained by Talreja (Talreja,1984) and Tay (Tay & Lim, 1993) was made. This analysis also used the $[0°/90°_3]_s$ glass/epoxy symmetric laminated with total thickness of 1,624mm. The results are presented in figure 9.

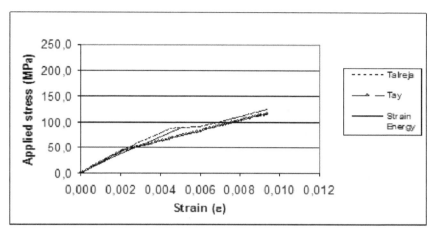

Fig. 9. Gl/Ep $[0°/90°_3]_s$ Laminated – Results comparison.

6. Conclusion

The present paper deals with damage composite laminate with transverse cracks in the matrix applying Continuous Damage Mechanics Theory, which was initially proposed by Kachanov (Kachanov, 1958) and than adapted by Allen (Allen et al., 1987a,b) for orthotropic laminated composites. This theory was also applied by Lim and Tay (Lim & Tay,1996) in laminates with transverse cracks to describe the stiffness loss of the structure. The adapted Damage Theory considers the mechanism associated to the transverse cracks through the internal state variables inside the constitutive relations based on the Continuous Damage Mechanics.

The theoretical model was implemented in a computational program, developed in FORTRAN language, based on the Modified Local Green's Function Method (MLGFM). The approximated solution was obtained by the MLGFM. The damage evolution model, originally developed for FEM, can be applied also to MLGFM without substantial changes in the original code.

In the presented results, it can be observed that the conventional criterions catch only the moment when the 90° layers no longer influences the stiffness of the laminated. Most of the criterions were able to determine the loss of stiffness. The strain energy criterion is able to evaluate the damage evolution, identifying the moment when the transverse cracks starts to affect the laminated rigidity. However, during the strain increase, the efficacy of the method to evaluate the stiffness loss decreases. Even so, as shown in the figure 8, the implemented code is able to denote, for all criterions, the stiffness loss in laminated composites when transverse cracks are formed in the matrix.

It is important to note that the actual stage of damage of a laminated plate depends on the historical of loading. As the micro cracks rise by quantity, length and opening, the external load must be applied step by step. A tolerance and a stopping strategy must be decisive for the accuracy and approximation of the true solution.

7. References

Allen, D. H.; Harris, C. E.; & Groves, S. E. (1987 a). A Thermo Mechanical Constitutive Theory for Elastic Composites with Distributed Damage – Part I: Theoretical Development, *Int. J. Solids Struct.*, Vol. 23, N. 9, pp 1301-1318.

Allen, D. H., Harris, C. E.; & Groves, S. E. (1987 b). A Thermo Mechanical Constitutive Theory for Elastic Composites with Distributed Damage – Part II: Application to Matrix Cracking in Laminated Composites, *Int. J. Solids Struct.*, Vol. 23, No. 9, pp. 1319-1338.

Allen, D. H.; Groves, S. E.; & Harris, C. E. (1988). A Cumulative Damage Model for Continuous Fiber Composite Laminates with Matrix Cracking and Interply Delaminations, *Composite Materials: Testing and Design (Eighth Conference)*, ASTM STP 972, J. D. Whitcomb, Ed., American Society for Testing and Materials, Philadelphia, pp. 57-80

Anderssen, R.; Gradin, P.; & Gustafson, C. G. (1998). Prediction of the stiffness degradation in cross-ply laminates due to transverse matrix cracking: an energy method approach, *Advanced Composite Materials.*, Vol 7, pp 325-346.

Barbieri, R.; Muñoz-Rojas, P. A.; Machado, R. D. (1998 a). Modified Local Green's Function Method (MLGFM) Part I. Mathematical background and formulation Eng. Anal. Bound. Elem. V. 22, pp. 141-151

Barbieri, R.; Muñoz-Rojas, P. A. (1998 b). Modified Local Green's Function Method (MLGFM) Part II. Application for accurate flux and traction evaluation Eng. Anal. Bound. Elem. V. 22, pp. 153-159

Barcellos, C. S.; Silva; L. H. M. (1987). Elastic Membrane Solution by a Modified Local Green's Function Method. *Proc. Int. Conference on Boundary Element Technology*. Ed. Brebbia, C. A , Venturini, W. S.. Southampton, U.K. of Applied Mechanics, V. 69, No. 3, pp. 145-159.

Cain, K.J.; Glinka, G.; Plumtree, A. (2003). Damage evolution in an off-axis unidirectional graphite bismaleimide composite loade in tension. *Composites: Parte A*, Vol 34, pp. 987-993.

Hashin, Z. (1987). Analysis of Orthogonally Cracked Laminates under Tension, *Transactions of The ASME*, Vol. 54, pp. 872-879.

Highsmith, A. L.; Reifsnider, K. L. (1982). Stiffness-Reduction Mechanisms in Composite Laminates, *Damage in Composite Materials*, ASTM STP775, Reifsnider K. L. – editor, American Society for Testing and Materials, pp 103-117.

Hull, D.; Clyne, T. W. (2000). An Introduction to Composite Materials. Cambridge University Press. ISBN: 0.521.38190-8.

Ji, F.S.; Dharani, L.R.; Mall, S. (1998). Analysis of transverse cracking in cross-ply composite laminates. Adv. Composite Mater., Vol 7, pp. 83-103.

Joffe, R. and Varna, J. (1999). Analytical Modeling of Stiffness Reduction in Symetric and Balanced Laminates due to Cracks in 90 Layers. *Composite Science and Technology.* Vol 59, pp. 1641-1652.

Kachanov, L. M. (1958) – On The Creep Fracture Time, Izv. Akad. Nauk SSR Otd. Tekhn. Nauk, N. 8, 26-31.

Kobayashi, S.; Ogihara S.; & Takeda N. (2000). Damage mechanics analysis for predicting mechanical behavior of general composite laminates containing transverse cracks. *Advanced Composite Materials,* Vol 9, pp. 363-375.

Lim, E. H.; Tay, T. E. (1996)- Stiffness Loss of Composite Laminates with Transverse Cracks under Mode I and Mode III Loading. *Int. J. Damage Mech.* Vol 5, pp. 190-215.

Machado, R.D.; Barcellos, C. S. (1992) - A First Modified Local Green's Function Method Approach to Orthotropic Laminated Plates. *Proceedings of CADCOMP-92, Computer Aided Design for Composite Materials Conference.* Ed. Brebbia, C. A , Newark, USA

Machado, R.D.; Abdalla Filho, J. E.; Silva, M. P. (2008). Stiffness loss of laminated composite plates with distributed damage by the modified local Green's function method. *Composite Structures,* Vol 84, pp. 220-227

Mendonça, P. T. R. (2005). Materiais Compostos e Estruturas Sanduíche, Projeto e Análise, Ed. Manole, São Paulo.

Ogihara, S.; Takeda, N.; Kobayashi, A. (1998). Analysis of stress and displacement fields in interlaminar-toughened composite with transverse cracks. *Adv. Composite Mater.,* Vol 7, No. 2, pp. 151-168.

Reddy, J. N.; Mechanics of Laminated Composite Plates: Theory and Analysis, Texas A&M University, College Station, Texas, CRC Press Inc.,1997.

Silberschimdt, V. (2005). Matrix cracking in cross-ply laminates: effect of randomness. *Composites: Part A,* Vol. 36, pp. 129-135

Talreja, R. (1984). Residual Stiffness Properties of Cracked Composite Laminates, *Proceedings of the Sixth Int. Conf. on Fracture,* ICFG, New Dalhi, India 4-10, December.

Talreja, R.; Boehler, J., P. (1990). Internal variable damage mechanics of composite materials, yielding, damage and failure of anisotropic solids. *Mechanical Engineering Publications,* EGF5, pp. 509-533.

Tay, T. E.; Lim, E. H. (1993) – Analysis of Stiffness Loss in Cross-ply Composite Laminates, *Composite Struct.,* Vol 25, pp.419-425.

Tay, T. E.; Lim, E. H. (1996) - Analysis of Composite Laminates with Transverse Cracks, *Composite Struct.* Vol 34, pp. 419-426.

Vasiliev, V. V.; Morozov, E. V. (2001). Mechanics and Analysis of Composite Materials. Elsevier Science Ltda. ISBN: 0-08-042702-2

Vejen, N.; Pyrz, R. (2002) – Transverse crack growth in glass/epoxy composites with exactly positioned long fibres. Part II: numerical. *Composites. Part B,* Vol. 33, pp. 279-290.

Wada, A.; Motogi, S.; Fukuda, T. (1999). Damage mechanics approach to nonlinear behavior
        of FRP laminates with cracking layers. *Adv. Composite Mater.*, Vol 8, No. 3, pp. 217-
        234

Permissions

The contributors of this book come from diverse backgrounds, making this book a truly international effort. This book will bring forth new frontiers with its revolutionizing research information and detailed analysis of the nascent developments around the world.

We would like to thank Dr. Yong X. Gan, for lending his expertise to make the book truly unique. He has played a crucial role in the development of this book. Without his invaluable contribution this book wouldn't have been possible. He has made vital efforts to compile up to date information on the varied aspects of this subject to make this book a valuable addition to the collection of many professionals and students.

This book was conceptualized with the vision of imparting up-to-date information and advanced data in this field. To ensure the same, a matchless editorial board was set up. Every individual on the board went through rigorous rounds of assessment to prove their worth. After which they invested a large part of their time researching and compiling the most relevant data for our readers. Conferences and sessions were held from time to time between the editorial board and the contributing authors to present the data in the most comprehensible form. The editorial team has worked tirelessly to provide valuable and valid information to help people across the globe.

Every chapter published in this book has been scrutinized by our experts. Their significance has been extensively debated. The topics covered herein carry significant findings which will fuel the growth of the discipline. They may even be implemented as practical applications or may be referred to as a beginning point for another development. Chapters in this book were first published by InTech; hereby published with permission under the Creative Commons Attribution License or equivalent.

The editorial board has been involved in producing this book since its inception. They have spent rigorous hours researching and exploring the diverse topics which have resulted in the successful publishing of this book. They have passed on their knowledge of decades through this book. To expedite this challenging task, the publisher supported the team at every step. A small team of assistant editors was also appointed to further simplify the editing procedure and attain best results for the readers.

Our editorial team has been hand-picked from every corner of the world. Their multi-ethnicity adds dynamic inputs to the discussions which result in innovative outcomes. These outcomes are then further discussed with the researchers and contributors who give their valuable feedback and opinion regarding the same. The feedback is then collaborated with the researches and they are edited in a comprehensive manner to aid the understanding of the subject.

Apart from the editorial board, the designing team has also invested a significant amount of their time in understanding the subject and creating the most relevant covers. They scrutinized every image to scout for the most suitable representation of the subject and create an appropriate cover for the book.

The publishing team has been involved in this book since its early stages. They were actively engaged in every process, be it collecting the data, connecting with the contributors or procuring relevant information. The team has been an ardent support to the editorial, designing and production team. Their endless efforts to recruit the best for this project, has resulted in the accomplishment of this book. They are a veteran in the field of academics and their pool of knowledge is as vast as their experience in printing. Their expertise and guidance has proved useful at every step. Their uncompromising quality standards have made this book an exceptional effort. Their encouragement from time to time has been an inspiration for everyone.

The publisher and the editorial board hope that this book will prove to be a valuable piece of knowledge for researchers, students, practitioners and scholars across the globe.

List of Contributors

J.F. Pommaret
CERMICS, Ecole Nationale des Ponts et Chaussées, France

Sushrut Vaidya and Jeong-Ho Kim
Department of Civil and Environmental Engineering, University of Connecticut, USA

Jianlin Liu
Department of Engineering Mechanics, China University of Petroleum, China

Quoc-Hung Nguyen and Ngoc-Diep Nguyen
Mechanical Faculty, Ho Chi Minh University of Industry, Vietnam

Chuan-Chiang Chen
Mechanical Engineering Department, California State Polytechnic University Pomona, USA

Xuesong Han
School of Mechanical Engineering, Tianjin University, P.R. China

Yong X. Gan
Department of Mechanical, Industrial and Manufacturing Engineering, College of Engineering, University of Toledo, USA

Roberto Dalledone Machado, Antonio Tassini Jr., Marcelo Pinto da Silva and Renato Barbieri
Pontifical Catholic University of Parana, Mechanical Engineering Graduate Program, Brazil

Printed in the USA
CPSIA information can be obtained
at www.ICGtesting.com
JSHW011809301024
72690JS00002B/16